D0713487

DATE DUE

DE _ 2 '91			
MR 6 '92 ~~RENEW~~			
MR 27 '92 4/10/09			

machine trades print reading

by

Michael Barsamian
Mechanical Drafting,
Blueprint Reading Instructor
Gateway Technical College
and
Richard Gizelbach
CNC Instructor
Gateway Technical College

South Holland, Illinois
THE GOODHEART-WILLCOX COMPANY, INC.
Publishers

STEPS IN LEARNING TO READ INDUSTRIAL BLUEPRINTS

Step 1 — Read the objectives at the beginning of each unit. They state what you will learn.

Step 2 — Read all the material in each unit and carefully study the directions before attempting to complete work problems.

Step 3 — Complete required sample work problems before attempting to answer industrial print reading questions.

Step 4 — Avoid the use of mechanical means of measuring the industrial prints. The prints are reproductions and are not true to scale. Blueprints should never be measured (scaled) for any dimension.

Step 5 — Use a straightedge or rule only on problems that call for sketching or view completion.

Library of Congress Catalog Card Number 86-347
International Standard Book Number 0-87006-575-0

23456789-86-098

Library of Congress Cataloging in Publication Data

Barsamian, Michael Allen.
  Machine trades print reading.

Includes index.
  1. Blueprints. 2. Machinery—Drawings.
  I. Gizelbach, Richard Allen. II. Title.
T379.B355 1986 604.2'5 86-347
ISBN 0-87006-575-0

CONTENTS

3

Introduction

MACHINE TRADES PRINT READING is designed to help you develop the basic skills required for visualizing and interpreting industrial blueprints. Each unit begins with competency based objectives you will achieve upon satisfactory completion of the unit.

The text consists of 12 units. The first four units give you the basics of print reading. They present information regarding the important concepts of visualizing shapes, line usage on drawings, and basic title block format.

Units 5 through 12 deal with actual working industrial prints. Each unit has specific technical information that has been applied to industrial prints.

The instructional materials, illustrations, and industrial prints in MACHINE TRADES PRINT READING are presented in a step-by-step process. The written material and illustrations in each unit contain the technical information needed to complete the industrial print reading questions. Each succeeding unit contains additional new material. Questions based on the industrial prints in each unit use the new material as well as learned material from earlier units.

Some of the industrial prints have certain changes or modifications from the original drawings to provide a variety of information that will aid your learning experiences.

Michael Barsamian

Richard Gizelbach

Unit 1

BLUEPRINTS AND BLUEPRINTING PROCESSES

After studying this unit, you will be able to:
- Explain why we need blueprints.
- Define the term "blueprint."
- Describe the photochemical blueprinting process.
- Describe the photocopying process.

When individuals create ideas, they form a mental image of that idea. Then, a "picture" is drawn to transfer that image and preserve it so that others may clearly understand that basic idea. These pictures or drawings are reproduced into what is commonly called "blueprints."

BLUEPRINTS are similar in nature to a map, which shows us how to get from one place to another. A blueprint, then, is a "map" that provides a route toward a finished product.

Certain types of information must appear on a blueprint. A craftworker uses this information to produce an actual object:

1. True size and true shape of the object or part.
2. All necessary size and location dimensions.
3. All symbols, notes, and special directions pertinent to manufacturing the object.

WHAT IS A BLUEPRINT?

A BLUEPRINT is a copy of an original drawing (master copy). The original drawing usually does not leave the engineering department because of possible damage to the original drawing sheet.

Often, a number of copies of an original drawing are needed or requested by individuals who are connected with manufacturing or processing a given part or a series of related parts. Therefore, a system

Fig. 1-1. This diazo printmaking machine uses seven fluorescent lamps, a 7 in. printing cylinder, and a pressure activator to produce prints. Finished prints are delivered front or rear; original and print are automatically separated. (Bruning)

of "blueprinting" was developed by which economical copies of that part drawing could be reproduced.

The blueprint serves as a valuable tool to industry. It is a major means of supplying necessary and technical information. With the sophisticated manufacturing processes now available, products are not only machined within the same plant, but often "jobbed out" to other vendors. In either case, the same specifications of the part must be maintained to produce a quality product.

The term "blueprint" usually is associated with the actual process of duplicating a drawing. The term originated long ago when all prints had a dark blue background and all lines and symbols were shown in white. The term has remained in use. Many drafters and others still say "blueprint" when they refer to any type of print.

THE DIAZO PROCESS

Today, blueprints (blue background/white lines) have been almost totally replaced by DIAZO reproductions. The diazo machine, Fig. 1-1, produces a black-line or blue-line image on a white background. In addition, various colored images can be produced on a variety of materials.

The diazo process is based, in principle, on the light sensitivity of diazo compounds. Diazo machines are available in dry, moist, and pressurized forms.

1. The moist form transfers an ammonia solution to the print to cause development. The print is delivered from the machine in a somewhat damp state.
2. The dry form of diazo printmaking uses ammonia vapor to develop the exposed copy.
3. The pressurized form utilizes a thin film of a special activator delivered under pressure to the exposed copy to complete the development. See Fig. 1-2.

MAKING PRINTS

There are various technical processes for obtaining a good quality "print." The two most common methods are PHOTOCHEMICAL and PHOTOCOPYING.

PHOTOCHEMICAL PROCESS

The primary photochemical process of printmaking involves a two-step method:
1. A piece of light-sensitive paper is covered by the original drawing. The paper and drawing are placed in the blueprint machine, Fig. 1-1. Light passes through the original drawing (which contains the part) and "burns" away the sensitized coating around the part outline. The remaining coated area, which has been protected by the lines of the part outline, turns blue when ex-

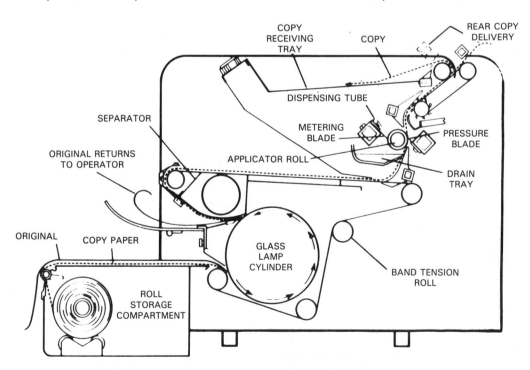

Fig. 1-2. To make a diazo print: Original and copy paper are fed into machine and routed around glass lamp to separator, where original is returned to operator. Copy paper continues past dispensing tube, applicator roll, and pressure blade. Print is deposited in copy receiving tray or delivered to rear copy delivery outlet. (Bruning)

posed in the developer. The background of the drawing generally will be white or a light shade of blue.

2. The sensitized print paper and the drawing are separated. Only the paper is passed through the developer section of the blueprint machine, Fig. 1-2. (The original drawing is placed to the side or put into its proper storage place to protect it from possible damage.) Through the use of ammonia vapor, the sensitized paper is transformed into a blue-line print that is actually dry to the touch.

The entire photochemical process takes only a few minutes in the blueprint machine.

Any changes to the part print must be performed on the original part drawing. Then, when the changes have been recorded, a new, updated blueprint can be issued to all departments or persons affected by the change.

The second most widely used photochemical method is the sepiaprint. The SEPIAPRINT PROCESS allows flexibility to create a master of the original drawing, or a print which can be corrected to produce new blue-line prints. Sepiaprints act as an "intermediate" between a second master copy and the original.

Like the blue-line prints, sepiaprints are sensitized to light and the ammonia developing process, Fig. 1-3. Sepiaprints, however, produce brown tone lines. The most significant difference between the sepiaprint and a blue-line print is the reversal of the original drawing to face the coated side.

The sepiaprint, then, is a reverse print, judging from the blue-line print standpoint. As this thin, translucent print is turned right-side-up, the image of the part is correct to the reader. Changes are made by means of a special correction fluid. When it is applied to the area to be corrected, it removes the brown tone. These prints are very versatile because you can actually make direct corrections on the sepia master. Then, more prints can be made showing the new change. See Fig. 1-3.

PHOTOCOPYING PROCESS

Photocopying of drawings has been used extensively throughout industry. PHOTOCOPYING is a reproduction process that involves the use of a special 35 mm camera and high contrast film. The original image size can be increased or decreased according to various drawing sizes.

The film negative of the part is attached to an aperture card containing the part number, description, and other important data. The aperture card can be previewed through a special viewfinder screen or directly produce a photoprint.

Aperture cards serve as a permanent record for new as well as old drawings. When changes occur to part drawings, new aperture cards must be made after each change notification. A trade term for this photocopying technique is called "microfilming."

LEARNING TO VISUALIZE

In order to interpret and understand blueprints, you need to develop the skill and ability to visualize (form a mental image of) an object on the basis of the drawing or the views of that object. This text will help you develop those basic skills.

Each unit is introduced by competency based statements. These are actual objectives you will achieve upon satisfactory completion of the unit.

COMPETENCIES are behavioral statements describing the knowledge, skill, and/or attitudes you should exhibit to achieve successful completion of an objective. Competencies define the learning outcome but need not include the conditions/limitations and criteria for acceptable performance.

Section I concentrates on the visualization of objects through the use of multiview presentation. Section II provides examples and explanations of information that appears on industry prints.

Metric units are omitted. Metric units do not affect the manner in which prints are read, but simply reflect mathematical conversions. A unit on

Fig. 1-3. This "whiteprinter" is designed to produce prints, sepias, or films. It utilizes four diazo lamps, a 5 in. pyrex glass cylinder, and an ammonia developing agent. No warm-up time is required. (Blu-Ray, Incorporated)

Geometric Tolerancing is included because the majority of industrial firms are utilizing this symbolism in describing machining procedures.

A heavy emphasis has been placed on the use of illustrations to help you develop the basic skills of visualization and simplify the learning process.

Actual industrial prints are featured to prepare you for the real conditions found in a machine shop.

In order to make the most of your study of MACHINE TRADES PRINT READING, follow the directions given in STEPS FOR LEARNING TO READ INDUSTRIAL BLUEPRINTS on page 2.

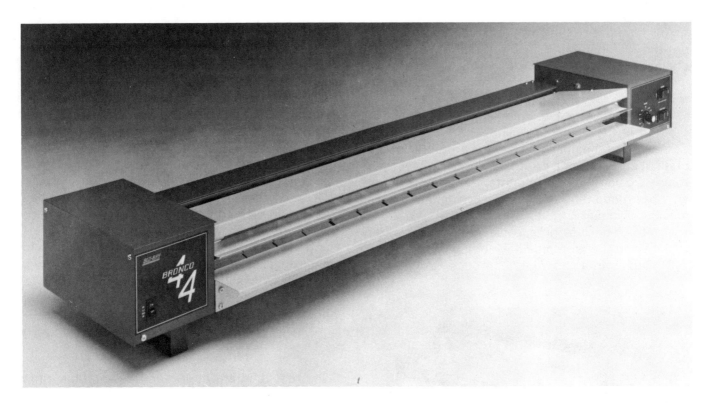

This self-contained reproduction station prints diazo materials at speeds up to 15 feet per minute. It provides a printing width of 44 5/8 inches with frontal print exit. A switch allows machine to be reversed to protect valuable originals. (Blu-Ray, Incorporated)

Unit 2

VISUALIZING SHAPES

After studying this unit, you will be able to:
☐ List the six primary views of an object.
☐ Select the appropriate dimensions of length, width, and height as they relate to various views.
☐ Draw a missing view of an object from two given views.

The key factor in all blueprint reading and drafting is the ability of the print reader or drafter to visualize separate views of an object on paper.

MULTIVIEW DRAWING

Understanding a multiview drawing is one of the most important single aspects of learning to read and interpret blueprints. This system of visualizing and describing a part is accepted throughout the engineering and manufacturing fields.

In MULTIVIEW DRAWING, every effort is made to show all necessary details of the shape and size of a part. Typically, the multiview drawing provides a means of visualizing a three-dimensional object (part) on a two-dimensional surface (drafting paper).

At first glance, the concept (idea) of three-dimensional visualization seems difficult to grasp. However, through the explanations and examples that follow, multiview representations will become more clear as you begin to visualize objects.

MULTIVIEW REPRESENTATION is a method of showing the shape of an object in two or more separate views projected at right angles to each other. PROJECTORS are straight lines of sight between views. Generally, the views are aligned with one another by means of ''perpendicular projectors.'' See Fig. 2-1. This method of drawing is known as ORTHOGRAPHIC PROJECTION. The terms ''multiview drawings'' and ''orthographic projection drawings'' are used interchangeably.

VIEWS IN A GLASS BOX

To further explore multiview representation, imagine that an object has been placed in a glass box. Three primary views are shown in Fig. 2-2. Top, front, and right side views have been transmitted to each glass surface by projectors. The projectors extend from the object to the flat surface of the glass, producing six different views of the object.

In Fig. 2-3, the surfaces are unfolded to show the six principal views: top, front, right side, left side, bottom, and rear. Visualize, then, how the glass box is being opened:
1. The front surface remains stationary.
2. The other surfaces hinge and rotate towards the front view surface.

As the six surfaces are unfolded, the projected views of the top, front, and right side become flat, like a sheet of drafting paper. In Fig. 2-4, all the view surfaces have been opened and carefully flattened to reveal their relative positions.

ARRANGEMENT OF VIEWS

Careful study should be given to the top view in Fig. 2-4, which is directly above and aligned through

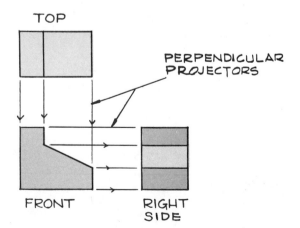

Fig. 2-1. Simple multiview drawing is done by orthographic projection. Three separate views are projected at right angles to each other.

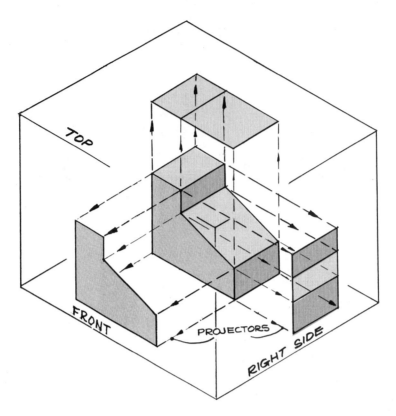

Fig. 2-2. An object is placed in a glass box, and three primary views are projected to glass surfaces.

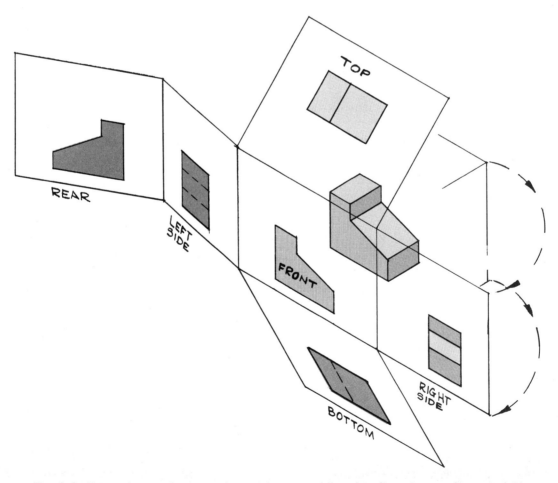

Fig. 2-3. Front view surface remains stationary, while other five view surfaces unfold.

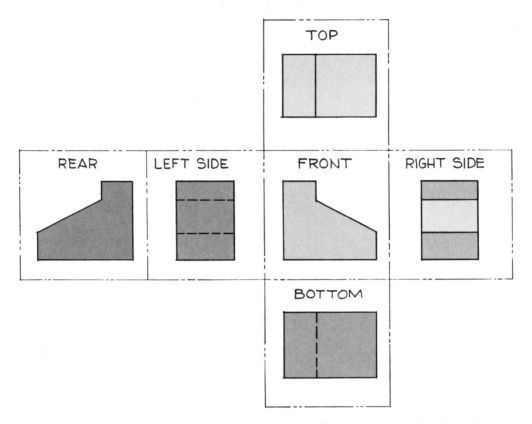

Fig. 2-4. Six principal view surfaces are flattened, like a two-dimensional drawing.

projectors with the front view. The right side view also has a direct relationship of projected lines to the right of the front view.

In actual work experience with industrial prints, all six views are rarely used. The most common combination of views is the top, front, and right side. Although the right side view is preferred, the left side view is used if it describes the object more clearly. Another exception occurs if the left side is more free of obstructing lines.

Fig. 2-5 shows a standard arrangement of top, front, and right side views, along with the basic special dimensions found on each view.

BASIC DIMENSIONS

The top view contains the length and width of the part. The front view will show the length and height; and, respectively, the right side view gives the height and width. Note that each view contains only two of the necessary special dimensions of the object.

Remember, in most cases, any single view of a machined part will not have sufficient information to describe the total size of the part. However, when any two of the three primary views (top, front, or right side) are given, the missing view can be solved. Refer back to Fig. 2-5 and carefully study the special dimensions on each view.

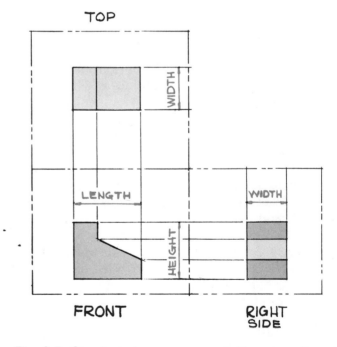

Fig. 2-5. Standard view arrangement: Top—Length and width. Front—Length and height. Right side—Height and width.

SURFACE REPRESENTATION

As noted, each view contains two basic measurements needed to complete the part. These two measurements are the basis for developing an individual surface. Fig. 2-6, for example, is a pictorial drawing of an object with front view A projected. A PICTORIAL DRAWING is a three-dimensional likeness of an object. In Fig. 2-6, the front projected surface A can be measured for true size and shape. note that projected surface A is identical to front view A on the pictorial drawing of the object.

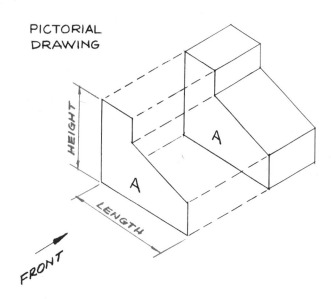

Fig. 2-6. Pictorial drawing shows front view projected and needing only two basic dimensions.

In Fig. 2-7, the top view differs from the front view in the representation of surface C. The pictorial drawing shows surface C as angular, while the projected surface C looks rectangular. The angular surface viewed from the top has been foreshortened through the projection process. Note that the width and length of the object have not been altered.

A similar procedure is used in Fig. 2-8 to obtain the right side view surface. This view presents surfaces D and E in true size and shape as projected from the pictorial drawing. Surface C becomes foreshortened again in the right side view. FORE-SHORTENED means contracted in the direction of depth. Also, the height and width have not changed in size from the projectors. With the completion of the right side view, this part would be drawn in the same layout pattern as in Fig. 2-5.

VISUALIZING BASIC SHAPES

Be sure to study and review the procedure for obtaining top, front, and right side views before attempting the more complicated objects. It is necessary and important to be able to visualize the basic shape of objects before reading other reference material on the print.

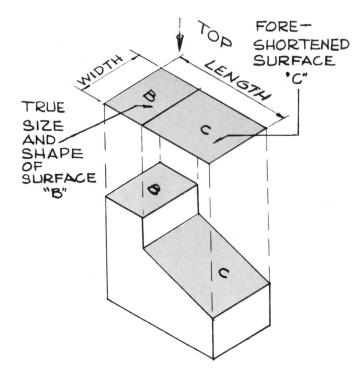

Fig. 2-7. In projecting top view of pictorial drawing, note that angular surface C is foreshortened.

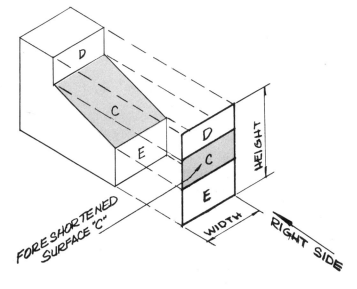

Fig. 2-8. Projected right side view of pictorial drawing also shows foreshortened angular surface C.

You will find that multiview drawings have levels of complexity that become evident as you learn more about the machining industry. Pay particular attention to the various forms or surfaces created by the lines of projection from the respective views. On most simple objects, the corners are drawn square (right angles) so as not to confuse the one who is visualizing the part. Fig. 2-9 shows a pictorial view of a basic ''L-shaped object.''

NOTE: The L-shape is not a practical machining project because most of the workpiece would be

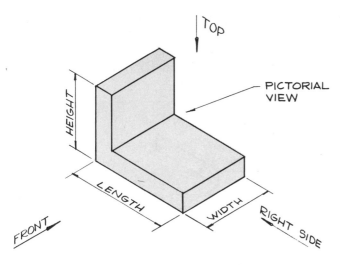

Fig. 2-9. Pictorial view of a basic L-shaped object is presented for ease of understanding visualization.

machined away in forming the L. However, the L-shape does lend itself well as a visualization exercise.

A MULTIVIEW APPLICATION

By applying the multiview method, the following representation of the L-shaped object will result. First, rotate the object into the front view position (in the same flat plane as paper). Note, as the front view is drawn, Fig. 2-10, the width cannot be seen.

PROJECTING THE VIEWS

Next, extend the projectors toward the top view and right side view positions. The projectors may

be extended as far as needed between each view. Then, locate the top view and right side views by transferring the width dimension. Use a miter angle (45°) to transfer the width from top to right side view, as shown in Fig. 2-11.

In many instances, simple objects may use the same top and right side views, but differ in front views. These interchangeable surfaces also use the same projectors without modifications. See Fig. 2-12.

Throughout this unit, you have been exposed to the basic method of visualizing multiview drawing representations. On the following pages, sample problems have been provided as a further guide to understanding multiview representations.

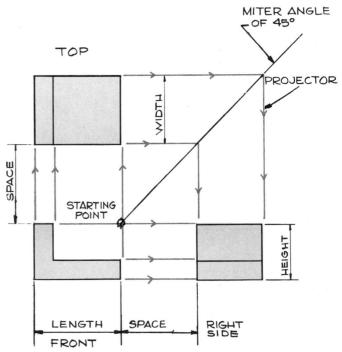

Fig. 2-11. When front and top views have been drawn, use extenders and miter angle to transfer width dimension to right side view.

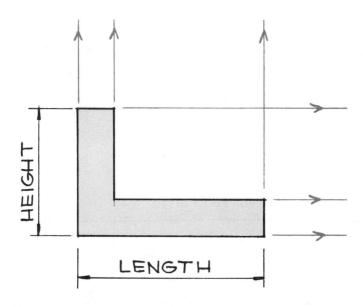

Fig. 2-10. L-shaped object is rotated into front view position. Note that right side cannot be seen.

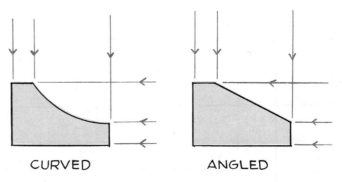

Fig. 2-12. Interchangeable surfaces on simple objects use same top and right side views. Difference is revealed in front views.

DIRECTIONS FOR SOLVING PROBLEMS

Complete the missing view in each problem. Use a ruler, machinist's scale, or other straightedge. 1. Label each view: top, front, and right side. 2. Include length, width, and height on each view.

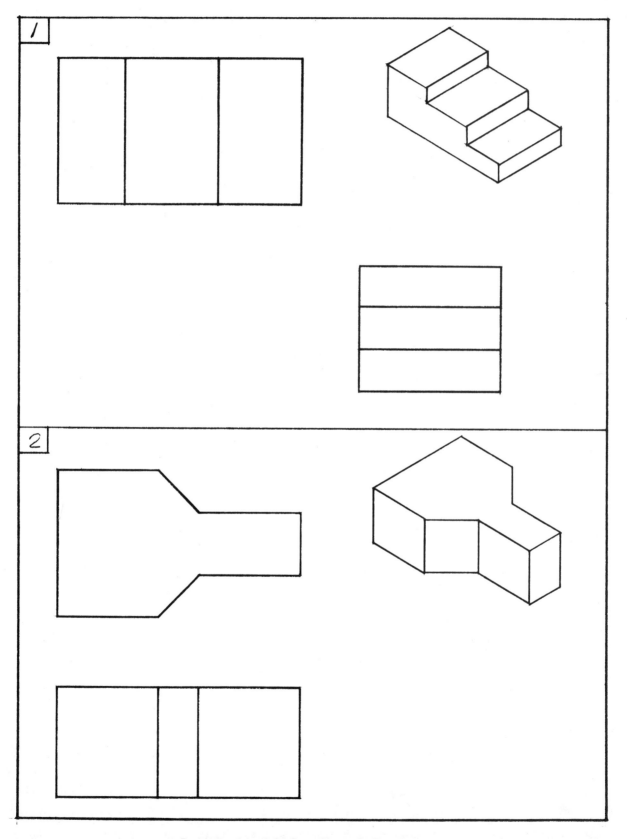

Problems 1 and 2. Complete missing view.

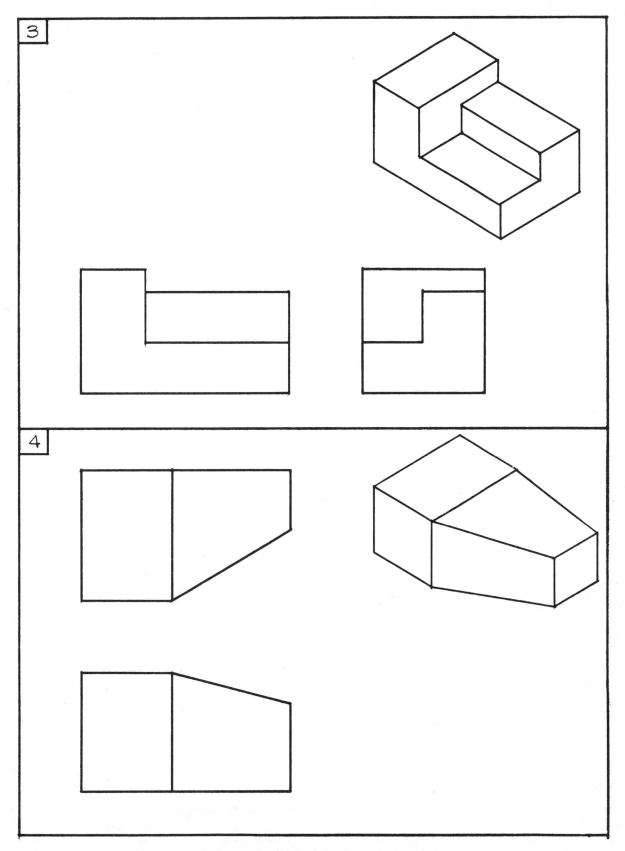

Problems 3 and 4. Complete missing view.

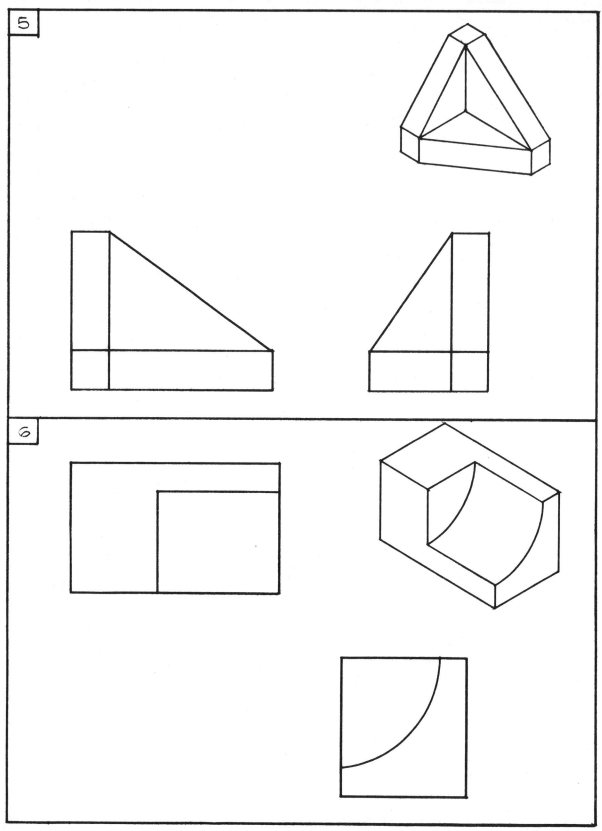

Problems 5 and 6. Complete missing view.

Unit 3

LINE USAGE ON DRAWINGS

After studying this unit, you will be able to:
☐ Identify various types of lines found on blueprints.
☐ Locate corresponding lines or surfaces in various views.

As new ideas are developed into salable products, more drawings and specialized machines are needed to complete the production of parts. As a result, a simple system of LINE USAGE (representation) on drawings has been developed as a "standard" for those involved with the production process.

TYPES OF LINES

Each newcomer in the field—or student studying drawings or prints—is expected to learn the use of these lines.

OBJECT LINE

Object lines are used to clearly show the outside border of the part or object. They also form surfaces within the borders of the part. Object lines are heavy, solid lines that stand out on a drawing. See Fig. 3-1.

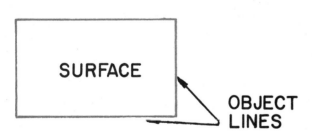

Fig. 3-1. Object lines.

HIDDEN LINE

Hidden lines show edges or surfaces which normally are not seen when viewing the part. The hidden line, Fig. 3-2, is easily identified by a series of thin-line dashes.

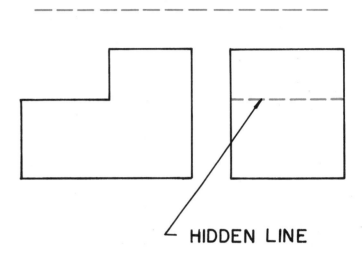

Fig. 3-2. Hidden line.

CENTERLINE

A centerline shows the location of the true center point of a (machined) hole or an axis of a part. It also may divide parts which are symmetrical (same on both sides). The centerline is a thin line, Fig. 3-3, easily noted by its arrangement of alternating long and single short dashes. This line symbol is most critical to the machinist during layout, machining, or when inspecting the part.

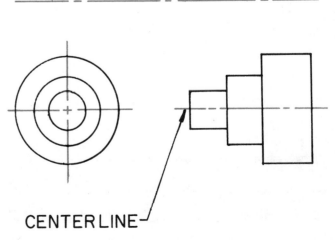

Fig. 3-3. Centerline.

SECTIONING LINES

Sectioning lines are used in the technique called "sectioning," which exposes the interior details of the part. Sectioning lines are evenly spaced, thin diagonal lines. See Fig. 3-4. As two or more parts are sectioned, the diagonal direction of one part changes. Different types of section lines are used to identify various types of material. See Unit 9 — Sectional Views.

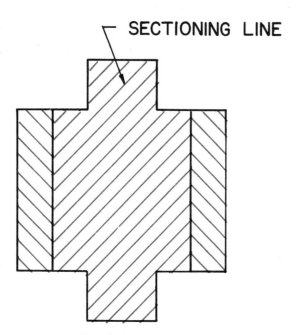

Fig. 3-4. Sectioning lines.

EXTENSION AND DIMENSION LINES

Extension and dimension lines are used together as a pair of thin lines forming a dimension, Fig. 3-5. The extension line extends away from the corners or surfaces of the part. The dimension line spans the distance between the extension lines with a number value and arrowheads.

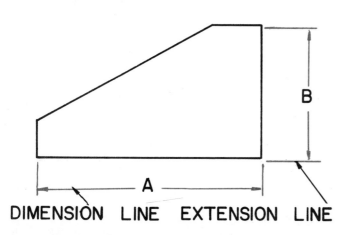

Fig. 3-5. Extension and dimensioning lines.

LEADER LINE

A leader line directs the reader to a specific operation or note vital to the machining of the part. The leader line couples a thin angular (slanted) line with an arrowhead. See Fig. 3-6.

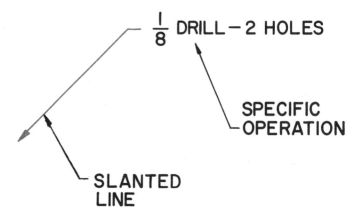

Fig. 3-6. Leader line.

CUTTING PLANE LINES

Cutting plane lines are directional indicators used in sectioning. They denote along which plane the part is cut, as well as from what direction the sectional view is taken. A cutting plane line is a thick, bold line with oversized arrows, Fig. 3-7. It is used with sectional views.

Fig. 3-7. Cutting plane line.

PHANTOM LINES

Phantom lines are used to indicate movement or motion. They also are used to show repeated details and/or extra material on a part before machining. Phantom lines are thin lines made up of one long dash followed by two short dashes. See Fig. 3-8.

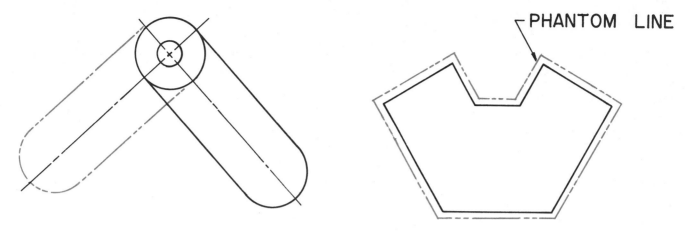

Fig. 3-8. Phantom line.

BREAK LINES

Break lines are used to "break" (shorten) long parts that are uniform in shape or cross section so that the size of the drawing can be reduced. Break lines are also used to provide clearer detail in view-ing the part or parts that lie directly below the removed part. Cylindrical and tubular objects call for curved break lines. Short breaks are shown by using thick, freehand lines, Fig. 3-9. Long breaks use thin, ruled lines with freehand zigzags. Break lines are used on both detail and assembly drawings.

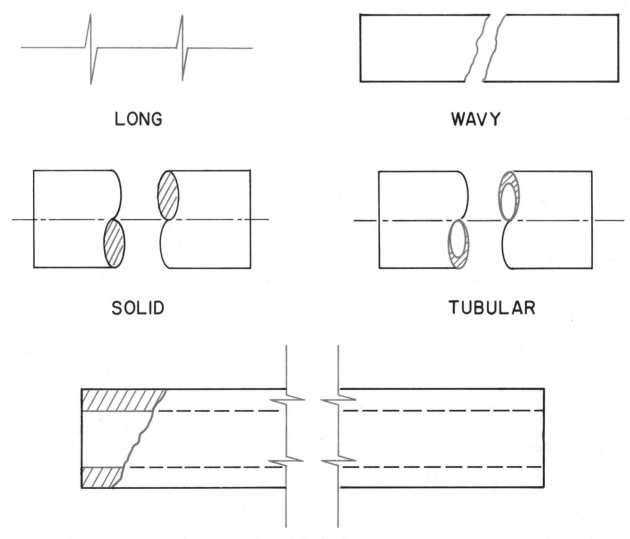

LONG

WAVY

SOLID

TUBULAR

Fig. 3-9. Break lines.

LINE IDENTIFICATION ON A PRINT

Fig. 3-10 shows line identification on a two-view drawing. Several types of lines described in this unit are identified and shown as they would appear on a typical print.

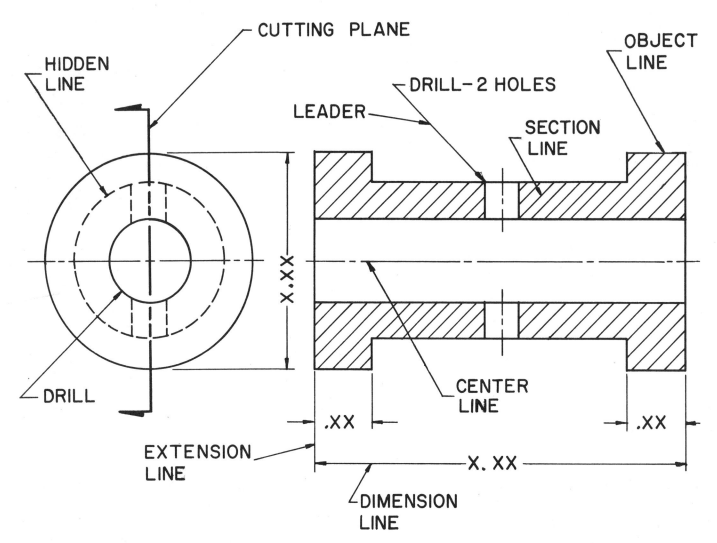

Fig. 3-10. Line identification.

LINE IDENTIFICATION

Given pictorial drawings with marked surfaces, identify the corresponding lines and surfaces on the top, front, and right side views for the following problems, Figs. 3-11, 3-12, and 3-13.

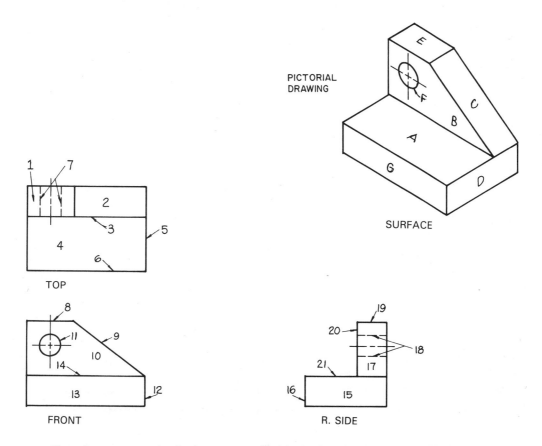

Place the correct number in the space provided for each surface of the pictorial drawing.

SURFACE	TOP	FRONT	R. SIDE
A	4	14	21
B	3	10	20
C	2	9	17
D	5	12	15
E	1	8	19
F	7	11	18
G	6	13	16

Fig. 3-11. Sample problem and answer grid.

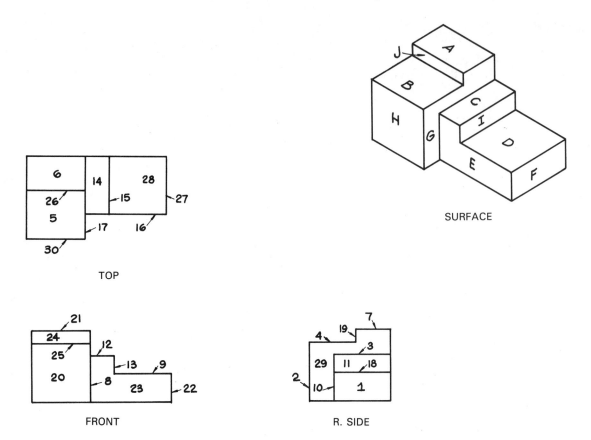

SURFACE

TOP

FRONT

R. SIDE

Place the correct number in the space provided for each surface of the pictorial drawing.

SURFACE	TOP	FRONT	R. SIDE
A			
B			
C			
D			
E			
F			
G			
H			
I			
J			

Fig. 3-12. Problem 1 and answer grid.

Line Usage on Drawings

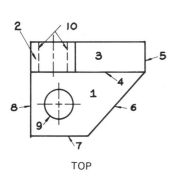

TOP

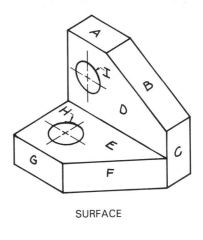

SURFACE

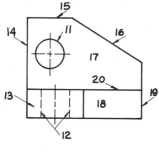

FRONT

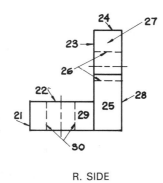

R. SIDE

Place the correct number in the space provided for each surface of the pictorial drawing.

SURFACE	TOP	FRONT	R. SIDE
A			
B			
C			
D			
E			
F			
G			
H			
I			

Fig. 3-13. Problem 2 and answer grid.

23

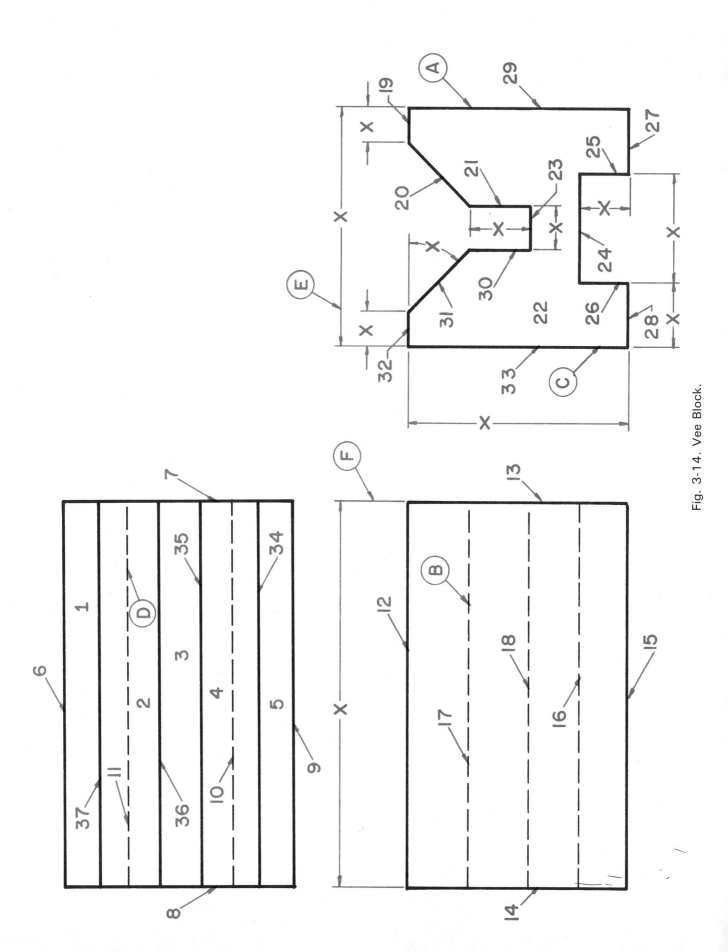

Fig. 3-14. Vee Block.

DIRECTIONS—QUIZ QUESTIONS

1. Familiarize yourself with the shape of the object.
2. Follow the hidden lines and centerlines from view to view.
3. Answer the questions for each of the drawings.

VEE BLOCK QUIZ

1. Line 25 in the side view is what line in the top view?   1. _____

2. How many surfaces are shown in the top view?   2. _____

3. Line 33 in the side view is what line in the top view?   3. _____

4. Line 18 in the front view is what surface in the top view?   4. _____

5. What type of line is line Ⓒ ?   5. _____

6. Surface 4 in the top view is what line in the side view?   6. _____

7. Line 16 in the front view is what line in the side view?   7. _____

8. Surface 22 in the side view is what line in the top view?   8. _____

9. Surface 22 in the side view is what line in the front view?   9. _____

10. Line 20 in the side view is what surface in the top view?   10. _____

11. Line 27 in the side view is what line in the front view?   11. _____

12. What type of line is line Ⓓ ?   12. _____

13. Surface 5 in the top view is what line in the side view?   13. _____

14. Line 23 in the side view is what surface in the top view?   14. _____

15. Line 10 in the top view is what line in the side view?   15. _____

16. Surface 1 in the top view is what line in the side view?   16. _____

17. What type of line is line Ⓐ ?   17. _____

18. Line 21 in the side view is what line in the top view?   18. _____

19. Line 14 in the front view is what line in the top view?   19. _____

20. Line 35 in the top view is what line in the side view?   20. _____

21. Line 29 in the side view is what line in the top view?   21. _____

22. What type of line is line Ⓔ ?   22. _____

23. What type of line is line Ⓑ ?   23. _____

24. Surface 1 in the top view is what line in the front view?   24. _____

25. What type of line is line Ⓕ ?   25. _____

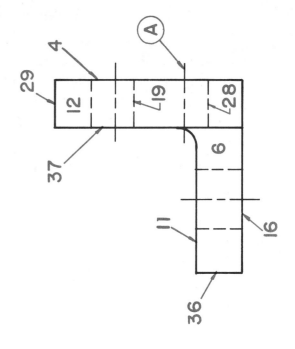

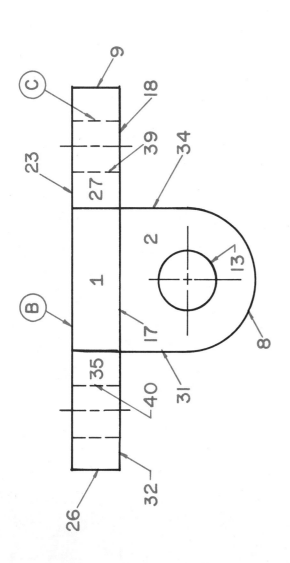

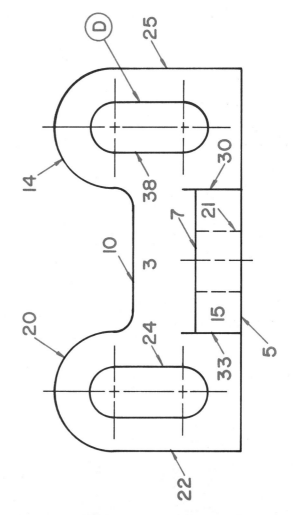

Fig. 3-15. Adjusting Bracket.

ADJUSTING BRACKET QUIZ

1. Line 7 in the front view is what surface in the top view?

2. Line 4 in the side view is what line in the top view?

3. Surface 1 in the top view is what line in the side view?

4. Line 29 in the side view is what surface in the top view?

5. Surface 1 in the top view is what line in the front view?

6. Line 9 in the top view is what line in the front view?

7. Surface 6 in the side view is what line in the front view?

8. Surface 6 in the side view is what line in the top view?

9. Line 24 in the front view is what line in the top view?

10. What kind of line is line (C) ?

11. Surface 3 in the front view is what line in the top view?

12. Line 13 in the top view is what line in the front view?

13. Surface 12 in the side view is what line in the top view?

14. Line 18 in the top view is what line in the side view?

15. What kind of line is line (B) ?

16. Line 20 in the front view is what surface in the top view?

17. Surface 27 in the top view is what line in the front view?

18. Line 38 in the front view is what line in the top view?

19. Line 20 in the front view is what line in the side view?

20. Line 22 in the front view is what line in the top view?

21. What kind of line is line (A) ?

22. Line 5 in the front view is what line in the side view?

23. Line 33 in the front view is what line in the top view?

24. What kind of line is line (D) ?

25. Surface 2 in the top view is what line in the side view?

1. _____

2. _____

3. _____

4. _____

5. _____

6. _____

7. _____

8. _____

9. _____

10. _____

11. _____

12. _____

13. _____

14. _____

15. _____

16. _____

17. _____

18. _____

19. _____

20. _____

21. _____

22. _____

23. _____

24. _____

25. _____

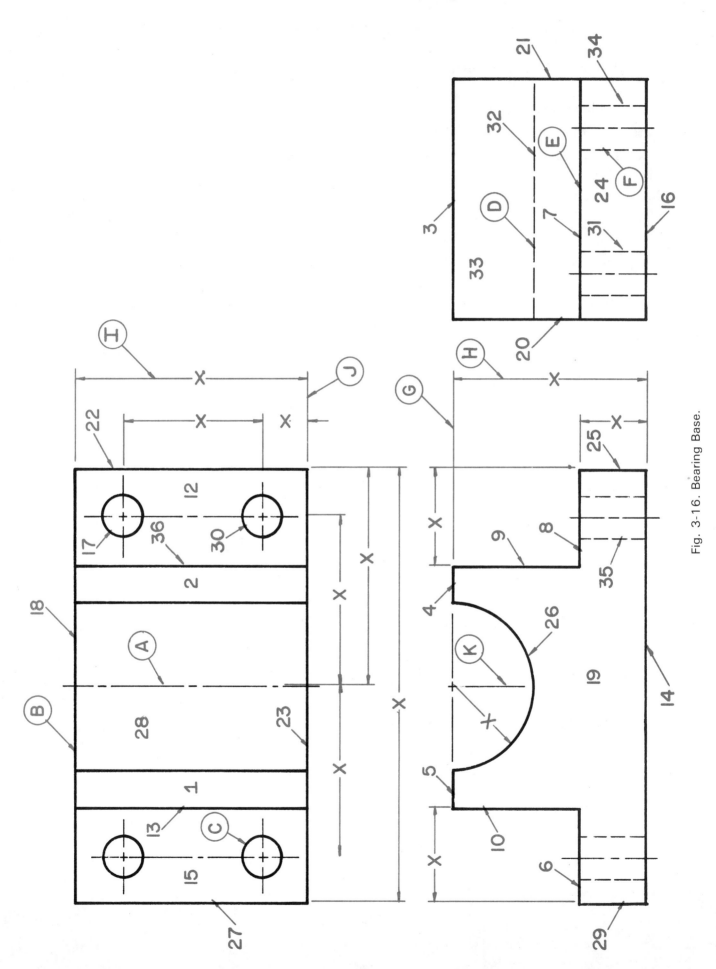

Fig. 3-16. Bearing Base.

BEARING BASE QUIZ

1. Line 23 is represented by what surface in the front view?

2. Line 9 in the front view is what line in the top view?

3. What type of line is (A) ?

4. Surface 28 is represented by what line in the side view?

5. Line 31 in the side view is what surface in the top view?

6. What type of line is (F) ?

7. Line 3 in the side view is what line in the front view?

8. Surface 19 in the front view is represented by what line in the side view?

9. Line 21 in the side view is what line in the top view?

10. What type of line is (B) ?

11. Line 22 is what line in the front view?

12. Surface 33 is what line in the front view?

13. Surface 24 is what line in the top view?

14. What type of line is (K) ?

15. Line 6 is what surface in the top view?

16. What type of line is (J) ?

17. Line 17 is what line in the side view?

18. What type of line is (H) ?

19. What type of line is (C) ?

20. Line 8 is what surface in the top view?

21. What type of line is (D) ?

22. Line 5 in the front view is what line in the side view?

23. What type of line is (E) ?

24. What type of line is (G) ?

25. What type of line is (I) ?

1. _____
2. _____
3. _____
4. _____
5. _____
6. _____
7. _____
8. _____
9. _____
10. _____
11. _____
12. _____
13. _____
14. _____
15. _____
16. _____
17. _____
18. _____
19. _____
20. _____
21. _____
22. _____
23. _____
24. _____
25. _____

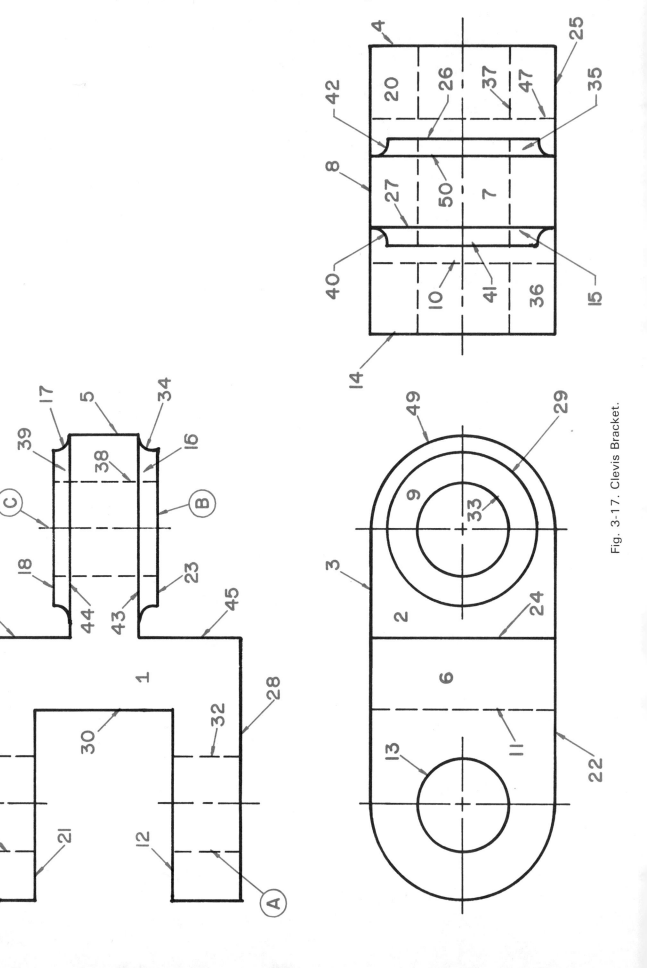

Fig. 3-17. Clevis Bracket.

CLEVIS BRACKET QUIZ

1. Surface 1 is denoted by what line in the side view?

2. Line 18 in the top view is represented by what line in the side view?

3. Surface 7 is represented by what line in the front view?

4. Line 12 in the top view denotes what line in the side view?

5. Surface 6 is represented by what line in the side view?

6. Surface 20 is denoted by what line in the top view?

7. Line 30 in the top view is represented by what line in the front view?

8. What kind of line is line (A) ?

9. Surface 9 represents what line in the top view?

10. Surface 9 is represented by what line in the side view?

11. Surface 36 represents what line in the front view?

12. Line 14 in the side view is represented by what line in the top view?

13. Line 13 in the front view is what line in the top view?

14. Line 37 is represented by what line in the front view?

15. Surface 15 in the side view is what surface in the top view?

16. Surface 7 in the side view represents what line in the top view?

17. What kind of line is line (B) ?

18. Line 37 is what line in the top view?

19. Line 43 in the top view is what line in the side view?

20. Surface 2 in the front view is represented by what line in the side view?

21. What kind of line is line (C) ?

22. Line 21 in the top view is what line in the side view?

23. Line 5 in the top view is what surface in the side view?

24. Line 29 denotes what line in the side view?

25. Surface 35 in the side view is what surface in the top view?

1. _____

2. _____

3. _____

4. _____

5. _____

6. _____

7. _____

8. _____

9. _____

10. _____

11. _____

12. _____

13. _____

14. _____

15. _____

16. _____

17. _____

18. _____

19. _____

20. _____

21. _____

22. _____

23. _____

24. _____

25. _____

Unit 4

BASIC TITLE BLOCK FORMAT

After studying this unit, you will be able to:

☐ Interpret information found in a title block, tolerance block, change block, and materials list.

☐ Determine tolerances on decimal, fractional, metric, and angular dimensions.

☐ Determine limits on decimal, fractional, metric, and angular dimensions.

☐ Define such terms as unilateral, bilateral, tolerance, low limit, high limit, nominal size, and basic size.

The major topics covered in this unit will include Title Blocks, Tolerance Blocks, Change Blocks, Materials List, Dimensions, Tolerances, Limits, and Notes.

TITLE BLOCK

The TITLE BLOCK is a boxed area containing general information about the part in the drawing. While it excludes the views, dimensions, and notes on any drawing or print, the title block represents an area of special interest for the print reader. Much of the information which is not drawn will be found in the title block format.

The basic title block formats generally used in industry are the STRIP STYLE and BLOCK STYLE. See Fig. 4-1. The type of data contained in each style is similar. However, the content and arrangement will vary since nearly every company has their own particular style of title block.

While it is impossible to present and explain all formats companies use, the examples given contain items that are typical of those that should be included in the title block.

NAME OF COMPANY

The NAME OF COMPANY area of the title block should include the name and location of the com-

pany. Often, the company's logo (symbol) may be part of this space. See Fig. 4-1.

DRAWING NUMBER

The DRAWING NUMBER serves as the identification number for filing purposes. Generally, most firms indicate the size of the drawing by a letter first; then a dash (—), followed by the actual number series. See A in Fig. 4-2. This system separates the various drawings by paper size, Fig. 4-3, for easy identification.

There are code systems, too, which use numbers first; then a letter, followed by other numbers. See B in Fig. 4-2. Some firms do not use any letters in their drawing number to identify drawing size, but merely use numbers to identify the drawing. See C in Fig. 4-2 and space A in Fig. 4-4.

As drawing sizes change, storage and identification becomes a major problem. To avoid this problem, most companies place the drawing number near the bottom right-hand corner of the title block format or in a separate space near the upper border line. When the print is correctly folded (large size prints), the title block with the company name and drawing number may be viewed for easy reference in the storage system.

Bear in mind that it is important to check the print (drawing) number before proceeding with specific work operations. There are occasions where drawings have been incorrectly processed with the work order because of similar drawing numbers.

Also, when checking the drawing number, pay particular interest to the drawing sheet size. Sometimes, the drawing size (letter code) and print are incorrectly matched.

PART NUMBER

The PART NUMBER is also a means for identifying a specific item or part. Part numbers are located

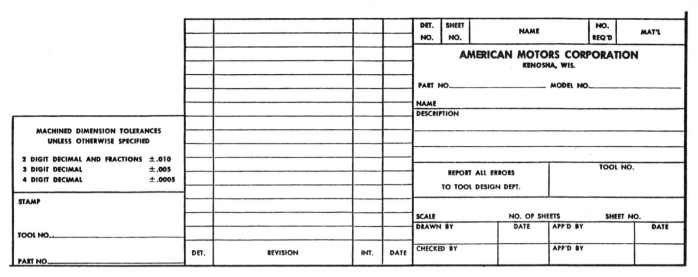

				DET. NO.	SHEET NO.	NAME	NO. REQ'D	MAT'L

AMERICAN MOTORS CORPORATION
KENOSHA, WIS.

PART NO._____ MODEL NO._____

NAME

DESCRIPTION

REPORT ALL ERRORS TO TOOL DESIGN DEPT.	TOOL NO.

SCALE		NO. OF SHEETS		SHEET NO.	
DRAWN BY	DATE	APP'D BY			DATE
CHECKED BY			APP'D BY		

MACHINED DIMENSION TOLERANCES
UNLESS OTHERWISE SPECIFIED

2 DIGIT DECIMAL AND FRACTIONS ±.010
3 DIGIT DECIMAL   ±.005
4 DIGIT DECIMAL   ±.0005

STAMP

TOOL NO._____

PART NO._____

DET.	REVISION	INT.	DATE

BLOCK STYLE

DESIGN APPROVAL	DATE	UNSPECIFIED TOLERANCES		This Drawing is the property of the Dumore Corporation. It must not be reproduced or copied without written permission.		PART NO.
DES.ENG.		.xxx ±.005				
MFG.ENG.		.xx ±.015				
Q.C.		.x ±.050				**DUMORE CORPORATION**
		Machine Finish ✓125 Max.				1300 17TH STREET, RACINE, WI 53403
SALES		Angles ±1°				
		Concentricity .005 TIR				
DO NOT SCALE DRAWING		Squareness .001 per in.				
		REMOVE ALL BURRS BREAK SHARP CORNERS		TITLE		

MATERIAL SPEC:		Req. per Piece	RECEIVED AS	DWG. BY	DATE	SCALE
		Pattern No.		CKD. BY	DATE	
					REVISIONS	

STRIP STYLE

Fig. 4-1. Title block formats, generally, are either block style or strip style.

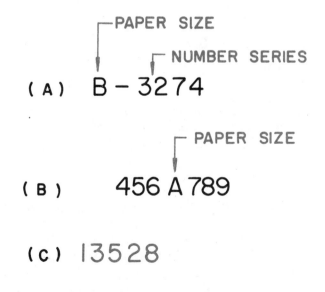

┌─ PAPER SIZE

      ┌─ NUMBER SERIES

(A)   B – 3274

      ┌─ PAPER SIZE

(B)   456 A 789

(C)   13528

Fig. 4-2. Drawing numbers are used for filing drawings and prints. Several popular drawing number code systems are shown.

SHEET SIZE	DIMENSIONS (INCHES)
A	8.5 x 11 or 9 x 12
B	11 x 17 or 12 x 18
C	17 x 22 or 18 x 24
D	22 x 34 or 24 x 36
E	34 x 44 or 36 x 48

Fig. 4-3. Drawing paper sizes are designated by letters. This chart lists sheet sizes and dimensions in inches.

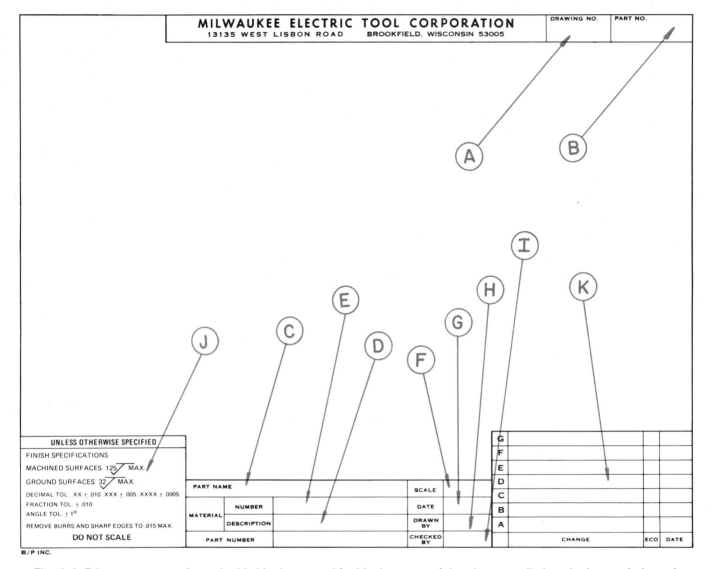

Fig. 4-4. Print components for main title block area and for blocks at top of drawing are called out by letters A through K. See text references for correct identification.

either in the main title block area or separately on the drawing. See space B in Fig. 4-4.

PART NAME

The PART NAME in the title block is the actual description or nomenclature assigned to the part. It may consist of a single word or several words. See space C in Fig. 4-4.

MATERIAL NUMBER

MATERIAL NUMBER is the number assigned by the company to identify the material being used. See space D in Fig. 4-4.

MATERIAL DESCRIPTION

MATERIAL DESCRIPTION denotes the specific

type of material required. It may specify the size of stock to be used. See space E in Fig. 4-4.

SCALE

SCALE indicates the ratio between the size of the object as it is drawn and the actual size of stock to be used. See space F in Fig. 4-4.

Typical Scales

1:1 or Full Scale	Print is actual size of object.
1:2, 1/2, or Half Size	Size of object appearing on print is one-half size of actual object.
1:4, 1/4, or Quarter Size	Size of object appearing on print is one-fourth size of object.

2X, 2:1, or Double Size — Size of object appearing on print is twice size of actual object.

NOTE: Scale only affects the size of the object on the print. Dimensions viewed on the print are always the actual sizes of the object regardless what scale is used.

DATE

DATE indicates when the print (drawing) was drawn by the drafter. See space G in Fig. 4-4.

DRAWN BY

DRAWN BY in the title block is followed by the initials of the drafter who produced the print (drawing). See space H in Fig. 4-4.

CHECKED BY

CHECKED BY is followed by the initials of the drafting department supervisor or manager, who checks the print to be sure that it is clear and accurate. See space I in Fig. 4-4.

UNLESS OTHERWISE SPECIFIED

FINISH SPECIFICATIONS
MACHINED SURFACES 125√ MAX.
GROUND SURFACES 32√ MAX.
DECIMAL TOL. .XX ± .010 .XXX ± .005 .XXXX ± .0005
FRACTION TOL. ± .010
ANGLE TOL. ± 1°
REMOVE BURRS AND SHARP EDGES TO .015 MAX.
DO NOT SCALE

UNSPECIFIED TOLERANCES	
.xxx	±.005
.xx	±.015
.x	±.050
Machine Finish	√125 Max.
Angles	±1°
Concentricity	.005 TIR
Squareness	.001 per in.

Fig. 4-5. Tolerance block gives general tolerance limits that apply, unless specific tolerance appears next to dimension on print.

TOLERANCE BLOCK

The TOLERANCE BLOCK indicates the general tolerance limits for one (.X), two (.XX), three (.XXX), and four (.XXXX) place decimal dimensions, or for fractional, metric, and angular dimensions. These limits in the tolerance block apply to decimal, fractional, metric, and angular dimensions unless the tolerance is specifically stated next to the dimension on the print.

Tolerances for the various "place dimensions" may vary from company to company. For example, the tolerance for a two-place (.XX) decimal dimension may be ± .010 inch for one company while another company may use ± .015 inch for two-place decimal limits.

SURFACE FINISH tolerances usually are included in this block. See space J Fig. 4-4. Manufactured parts usually require certain surface finishes. These surface finishes are designated by numbers which basically describe how smooth the surfaces must appear. As the values of these numbers decrease, the surface conditions improve. For example: a "125" specified finish is a "machined" finish. A "32" specified finish is a "ground" finish. Surface finish numbers can vary anywhere from 500 to 1. See Fig. 4-5.

CHANGE BLOCK

The CHANGE BLOCK or REVISION BLOCK is a record of changes made to the original drawing. It may be a structured block of information, as shown at K in Fig. 4-4. Or, it may be a simple block heading, Fig. 4-6, under which necessary information regarding revisions is added.

DRAWING REVISIONS are made to improve part design, reduce costs, clarify instructions, change dimensions, correct errors, and change manufacturing procedures.

There are many different format designs for change blocks. The examples given are just a few of the many variations that exist.

A typical structured change block might contain the information shown in Fig. 4-7:

Column A (SYM) Sequence of changes as they occur. A symbol (SYM) is shown next to actual change on print as well as being recorded in change block. The symbol usually is a letter of the alphabet, beginning with the letter A.

Column B (E.C.R. NO.) The Engineering Change Revision Number identifies the document which contains the record data, notes, and/or sketches that call for changes on the print. Changes could include: dimensions, materials, processes, notes, features, finishes, etc.

Column C (DATE) Date engineering change was entered.

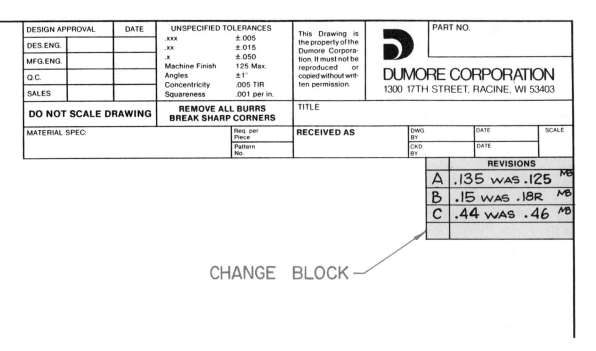

DESIGN APPROVAL	DATE	UNSPECIFIED TOLERANCES		This Drawing is the property of the Dumore Corporation. It must not be reproduced or copied without written permission.		PART NO.		
DES.ENG.		.xxx	±.005					
MFG.ENG.		.xx	±.015					
		.x	±.050		**DUMORE CORPORATION**			
Q.C.		Machine Finish	125 Max.					
		Angles	±1°		1300 17TH STREET, RACINE, WI 53403			
SALES		Concentricity	.005 TIR					
		Squareness	.001 per in.					
DO NOT SCALE DRAWING		**REMOVE ALL BURRS BREAK SHARP CORNERS**		TITLE				
MATERIAL SPEC:		Req. per Piece		**RECEIVED AS**	DWG. BY	DATE		SCALE
		Pattern No.			CKD. BY	DATE		

	REVISIONS	
A	.135 WAS .125	MB
B	.15 WAS .18R	MB
C	.44 WAS .46	MB

CHANGE BLOCK —

Fig. 4-6. Change block, or revision block, may be a simple listing of changes made to original drawing.

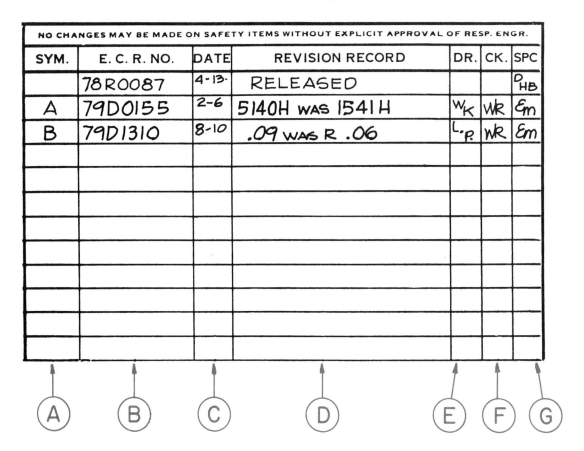

NO CHANGES MAY BE MADE ON SAFETY ITEMS WITHOUT EXPLICIT APPROVAL OF RESP. ENGR.

SYM.	E.C.R. NO.	DATE	REVISION RECORD	DR.	CK.	SPC
	78R0087	4-13-	RELEASED			D HB
A	79D0155	2-6	5140H WAS 1541H	W K	WR	Em
B	79D1310	8-10	.09 WAS R .06	L P	WR	Em

Ⓐ   Ⓑ   Ⓒ   Ⓓ   Ⓔ Ⓕ Ⓖ

Fig. 4-7. Change block may be "structured" to provide more columns and additional detail. See text references.

Column D (REVISION RECORD) The revision record is an explanation of the changes made. Sometimes the terms DESCRIPTION or CHANGE are used to head this column.

Columns E, F, and G (DR, CK, SPC) Contains authorizing initials of individuals responsible for drawing the change, checking the change, and recording the change in the parts book. DR means drawn. CK means checked. SPC means specifications. The SPC column bears the initials of an individual in the records department who enters the change in a parts book. The SPC column is special and is not likely found on other company revision blocks.

MATERIALS LIST

The MATERIALS LIST may also be called a "parts list" or a "bill of materials," or a "parts schedule." The materials list is used primarily on assembly type drawings which show more than a single part. As company needs vary, the materials list also changes.

The column headings found in the materials list block include Item, Part Number, Part Name, Required, and Material. See Fig. 4-8.

description of the part name consists of one or several words.

REQUIRED

The required column contains the actual number of parts required for the assembly.

MATERIAL

The material column contains a commercial description of material used.

DIMENSIONS

A DIMENSION is a definite measurement value given on a print such as length, width, height, radius, etc. Dimensions on a print can appear as fractional, decimal, angular, and metric values.

ITEM	PART NUMBER	PART NAME	REQ.	MATERIAL
4	3964	SLIDE BASE	1	SAE 06
3	5718	PLATE	1	1020 C.R.S.
2	6263	ROLLER	4	SAE 06
1	918	PIN	4	SAE 06
MATERIALS LIST				

Fig. 4-8. Materials list contains description and quantity of specific parts shown in an assembly drawing.

ITEM

The item is a number which directs the print reader to a specific part on the assembly drawing. Items are easily noted by a leader line and number within a circle.

PART NUMBER

Part identification refers to the actual manufacturer's part number. The parts (by number) which are purchased from other firms are also listed.

PART NAME

The part name or description indicates the actual name of that item in the assembly. Usually, the

FRACTIONAL DIMENSIONS

Fractional dimensions found on a print include 64ths, 32nds, 8ths, 4ths, and halves. A typical fractional dimension would look like Fig. 4-9.

DECIMAL DIMENSIONS

Workpieces that require more exact and closer dimensions usually have decimal dimensions. Decimal dimensions are parts of an inch that are commonly expressed as .10 (tenths), .01 (hundredths), .001 (thousandths), and .0001 (ten-thousandths) of an inch. The great majority of prints use decimal dimensions expressed in thousandths and ten-thousandths of an inch. Typical decimal dimensions appear in Fig. 4-10.

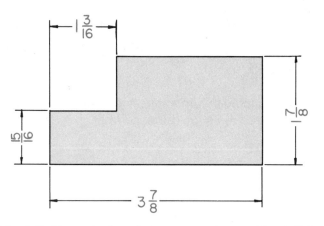

Fig. 4-9. Dimensioning of a part on a print provides definite measurement values. Fractional dimensioning is shown.

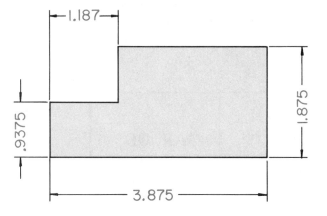

Fig. 4-10. Decimal dimensioning is used on prints where workpieces require more precise measurement.

ANGULAR DIMENSION

Angle size on a print is expressed in degrees. A degree (symbol °) is 1/360th of a circle. A degree can be broken down into minutes and seconds. There are 60 minutes (symbol ') to one degree and 60 seconds (symbol '') to one minute.

      Circle = 360 degrees (360°)
      Degree = 60 minutes (60')
      Minute = 60 seconds (60'')

A typical angular dimension would appear on a print as shown in Fig. 4-11.

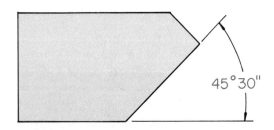

Fig. 4-11. Angular dimensions are indicated in degrees (°), in minutes ('), and in seconds ('').

METRIC DIMENSIONS

Blueprints having metric dimensions are in greater use due to expanded economic activity with countries that use the metric system in manufacturing. Metric dimensions on a print are stated in millimeters (mm), Fig. 4-12. Prints can be converted from the English system to the metric system on the basis of one inch equaling 25.4 millimeters.

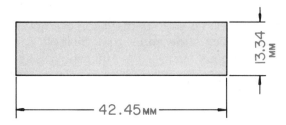

Fig. 4-12. Typical use of metric dimensioning, in millimeters, is illustrated.

DUAL DIMENSION

Companies making a slow transition toward the use of metric dimensioning, and/or whose product or products are used and manufactured in world markets, employ what is called ''dual dimensioning.'' Dual dimensioning is a method that states values in both the decimal inch system and its metric equivalent.

Several methods of dual dimensioning are in frequent use. Dual dimensions are often placed above and below dimension lines as shown in view A in Fig. 4-13. Or, dual dimensions are placed next to each other, using a slash as in View B. On some drawings, brackets, [ ], or parentheses, ( ), are used to separate each system as shown in views C and D.

TABULAR DIMENSIONS

When a company manufactures a family of parts or assemblies which are exactly alike in shape except for different dimensions, tabular dimensioning is used on the prints (drawings). Letters or numbers are used instead of dimensions to denote sizes, and a table is placed on the print to supply the corresponding values for the different parts. Refer to Fig. 4-14.

Tabular dimensioning is also used on prints which contain a large amount of repetitive features, such as holes. This method is used to avoid difficulty and confusion in reading the print because of all the extension and dimension lines needed. Instead, holes are labeled with a letter or number, and the location of the hole is provided on a chart based on a

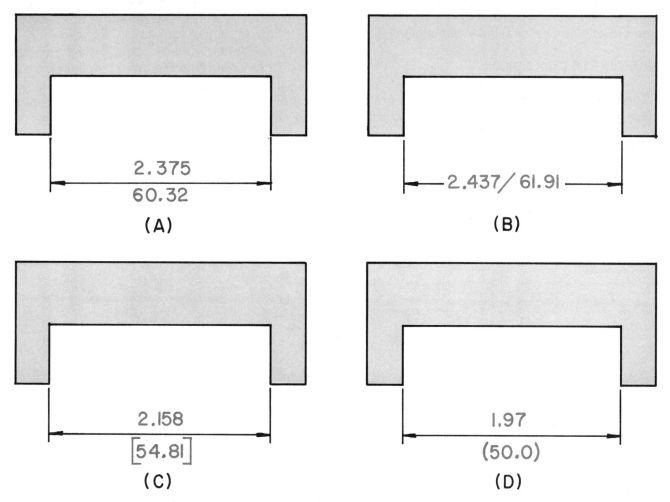

(A)

(B)

(C)

(D)

Fig. 4-13. Dual dimensioning gives measurements in both English and metric values.
See text references.

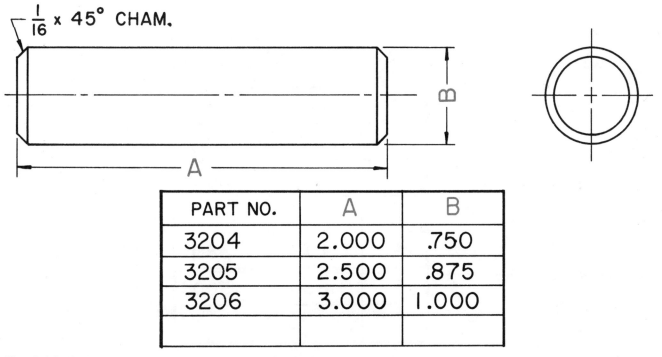

PART NO.	A	B
3204	2.000	.750
3205	2.500	.875
3206	3.000	1.000

Fig. 4-14. With tabular dimensioning, code letters or numbers appear on print instead of numerical values. A chart correlates these codes and values.

X-Y coordinate axis system. Fig. 4-15 provides an example of this use of tabular dimensioning.

COORDINATE DIMENSIONS

Coordinate dimensioning is used on prints that would require many dimensions and extension lines.

To help clarify and avoid difficulty in interpreting the print, distances are taken from vertical and horizontal datum (line or surface used as a reference) lines or surfaces to centerlines representing the location or locations desired. Refer to Fig. 4-16. Coordinate dimensioning is sometimes referred to as "arrowless" or "datum" dimensioning.

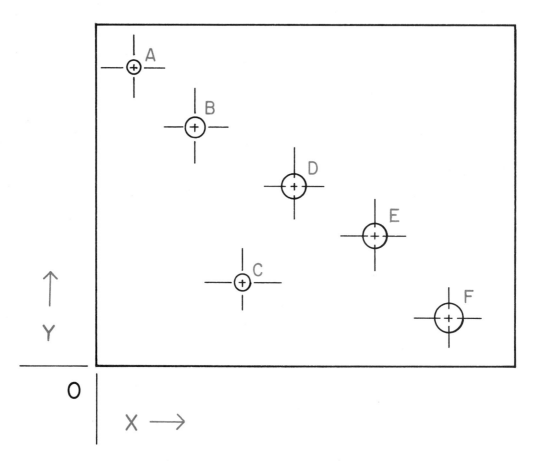

HOLE NO.	X	Y
A	.480	3.120
B	1.080	2.500
C	1.600	.800
D	2.060	1.900
E	3.100	1.380
F	3.820	.500

Fig. 4-15. Tabular dimensioning is used when dimensioning repetitious features would cause confusion. In this example, holes are coded by letters, and a chart correlates these letters and dimensions.

Basic Title Block Format

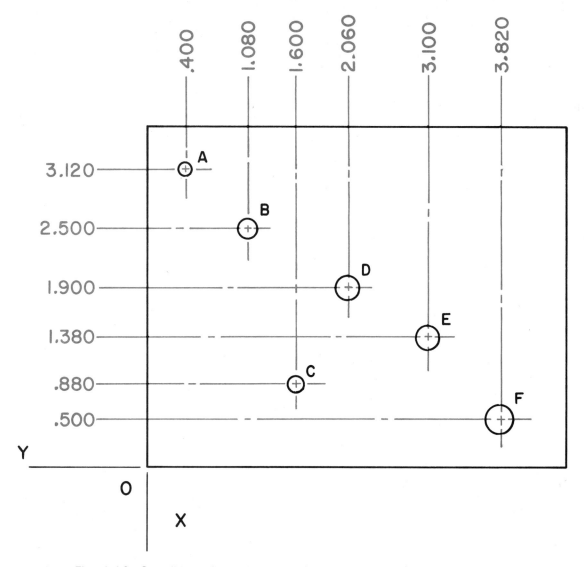

Fig. 4-16. Coordinate dimensioning is frequently used on prints of parts to be machined by numerical control. See text explanation.

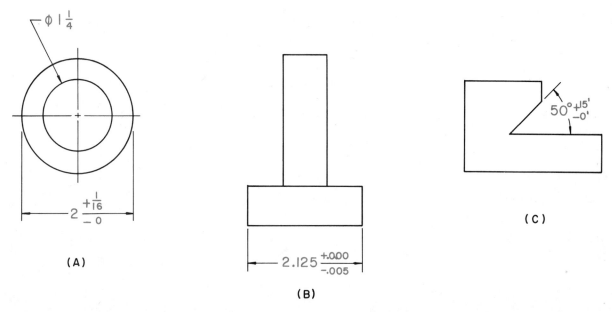

Fig. 4-17. Unilateral tolerances indicate variation permitted from basic dimensions in one direction only (either + or −).

TOLERANCES

TOLERANCE is the entire amount of variation permitted in the size of a part or object. Each dimension on a print has a tolerance applied to it to compensate for deviations that may occur during manufacturing. Tolerances are extremely important in providing sufficient size differences to maintain the interchangeability of parts and the proper assembly of units.

UNILATERAL TOLERANCES

Unilateral tolerance is variance from the basic dimension in only one direction, either plus (+) or minus (−). See Fig. 4-17, A, B, and C.

BILATERAL TOLERANCES

Bilateral tolerance is variance from the basic dimension in both directions: plus (+) and minus (−) dimensions. See Fig. 4-18, A, B, and C.

UNSPECIFIED TOLERANCES

Unspecified tolerances for fractional, decimal, angular, and metric dimensions are expressed through the use of a note as shown in Fig. 4-19.

SPECIFIED FRACTIONAL TOLERANCES

Specified fractional tolerances appear on a print as shown in Fig. 4-20.

SPECIFIED DECIMAL TOLERANCES

Specified decimal tolerances on a print can be expressed in several ways. See Fig. 4-21.

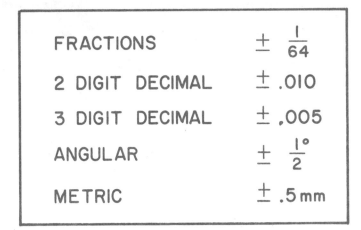

Fig. 4-19. Unspecified tolerances appear in a note on a print. They apply to dimensions in general.

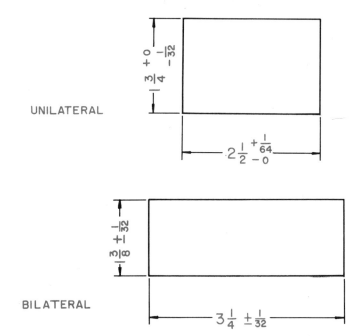

Fig. 4-20. Specified fractional tolerances are used in conjunction with a particular dimension.

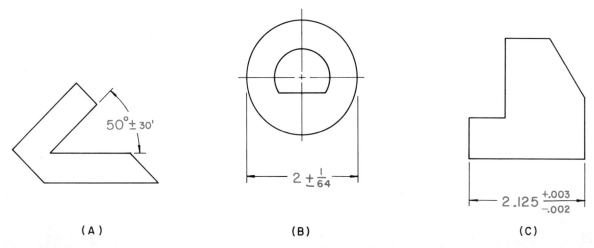

(A) (B) (C)

Fig. 4-18. Bilateral tolerances indicate variation permitted from basic dimensions in both directions (+ and −).

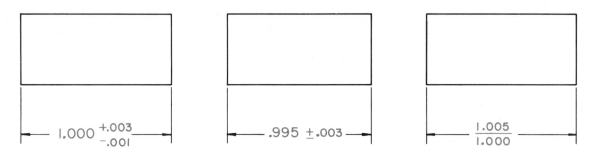

Fig. 4-21. Several ways of expressing specified decimal tolerances.

SPECIFIED ANGULAR TOLERANCES

Specified tolerance for angular dimensions appear as those shown in Fig. 4-22.

SPECIFIED METRIC TOLERANCES

Specified metric tolerances are found on a print as examples shown in Fig. 4-23.

LIMITS

LIMITS are the maximum and minimum sizes allowed for a specific dimension. Success in manufacturing relies on the ability to duplicate numerous parts to near identical specifications for assembly purposes.

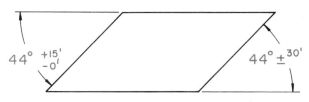

Fig. 4-22. Specified angular tolerances are used in conjunction with basic degree dimensions.

HIGH LIMIT

The high (upper) limit is the maximum allowed size given to a dimension. Under this condition, the high limit is the largest dimension used for manufacturing. See Fig. 4-24.

LOW LIMIT

The low (lower) limit is the minimum acceptable size dimension for the part. Likewise, the low limit will be used for some manufacturing processes, Fig. 4-24. Also, the numbers between 1.650 and

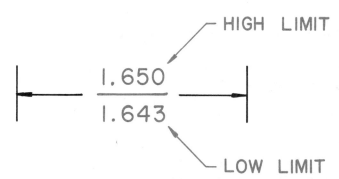

Fig. 4-24. Limit dimensions are maximum and minimum acceptable sizes for a part on a print.

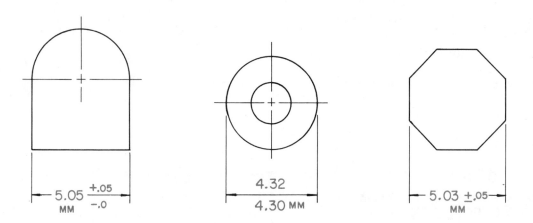

Fig. 4-23. Several ways of expressing specified metric tolerances.

43

1.643 are permissible dimensions. When two dimensions are used to show the limits, the high limit is placed above the low limit. In this case, the variation (tolerance) between the high limit (1.650) and the low limit (1.643) is .007.

In dealing with limit dimensions, certain terms relating to size should first be understood.

NOMINAL SIZE

Nominal size is the general size used for the purpose of identification of a part. It may or may not be the actual (measured) size of the part. For example, a 3/4 — 10 UNC bolt has a 3/4 inch nominal size diameter, however the actual diameter size may vary from .750 to .7288.

BASIC SIZE

Basic size is the theoretical size of a part from which limits for that dimension are set by the application of allowances and tolerances. For example, a shaft may have a basic diameter of 1.499 inches, but a maximum variation of .001 inch may be allowed.

The minimum size of the hole for this shaft should be of basic size in most cases because of the standard tools used in the machining of these holes. Fig. 4-25 shows the application of basic size regarding the assembly of a shaft and a bearing.

NOTES

Notes on a print provide added data and information not found in the views but needed for the part. They provide data not found in the title, tolerance, or revision blocks.

Notes provide information such as heat-treat directions and specifications, machining directions, finish requirements, measuring requirements, etc.

Notes should be read first before studying the views of the part because they may advise you of certain requirements or restrictions regarding the part. The word "note" sometimes precedes information it is furnishing.

Some typical notes found on prints could include:

FINISH: BLACK OXIDE
    FINISH THIS SIDE WORKING FROM
    COMPLETED OPPOSITE END

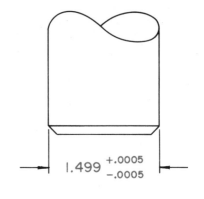

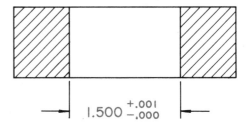

BASIC SHAFT DIA. 1.499
BASIC BEARING DIA. 1.500

Fig. 4-25. Basic size of a part is theoretical size from which allowances and tolerances are set.

BREAK ALL SHARP CORNERS
    NO DIE PARTING MARKS
    PERMISSIBLE ON THIS SURFACE

NOTE:
    WHEN GRINDING .3762/.3757 DIA.
    USE MICROMETER NOT AIR SNAP GAUGE
    HAVE SLIGHT TAPER TOWARDS PINION

NOTE:
    HEAT TREAT PER ES-12-1001
    TEMPER PER ES-12-3000 R 39-43
    ES NUMBER FOR IN-HOUSE USE

MINIMUM DEPTH OF 5/8 DRILL TO
CLEAR SPINDLE END

NO SHARP EDGE ON 4 SQUARE CORNERS

#3 GA. (.2391)
HOT ROLLED,
PICKLED AND
OILED SHEET
STEEL

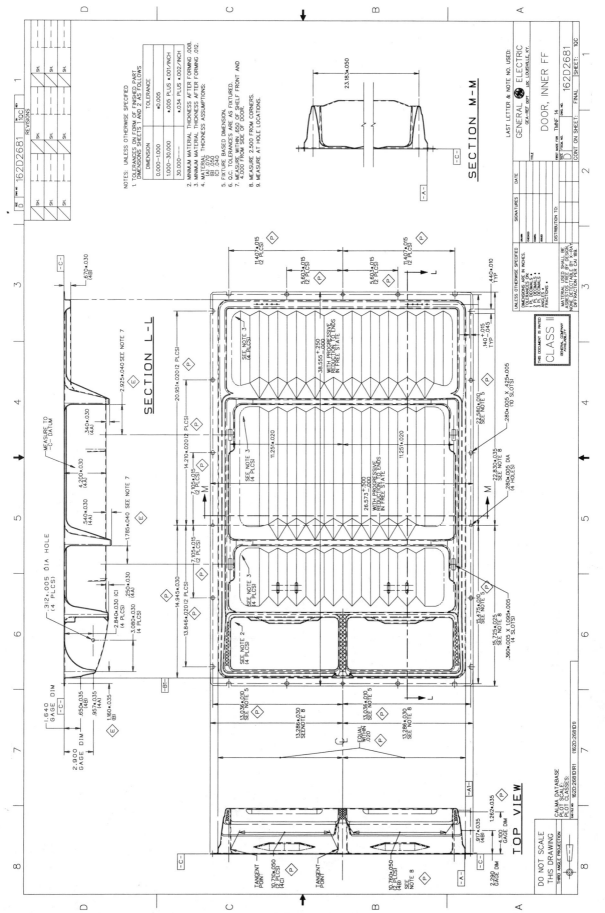

You will find that many prints were generated on a computer-aided drafting and design system. (General Electric)

45

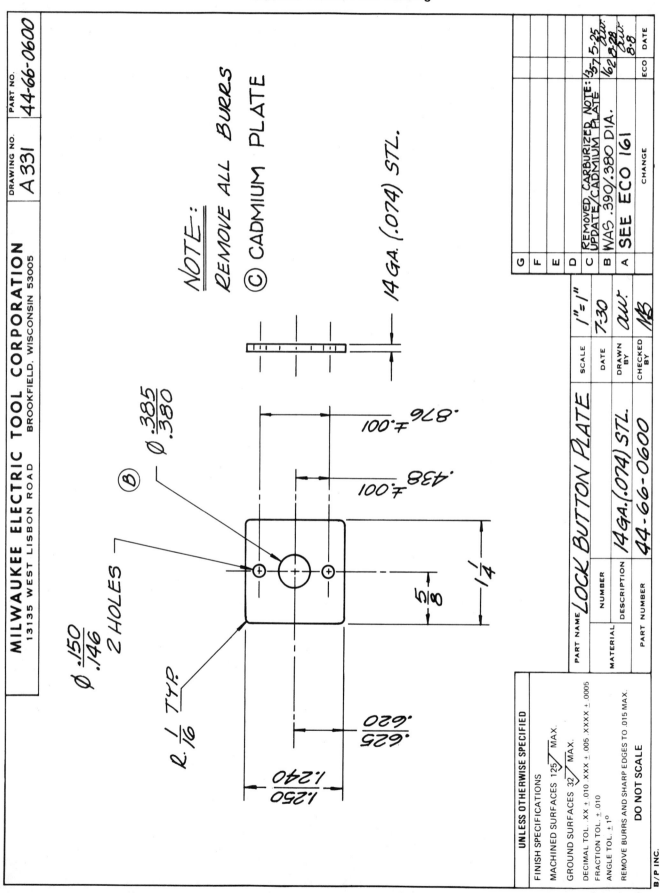

Fig. 4-26. Lock Button Plate.

DIRECTIONS — QUIZ QUESTIONS

1. The industrial prints in this section will test your print reading ability.
2. Study the views, dimensions, title block, and notes in Figs. 4-26 and 4-27.
3. Read the quiz questions, refer to the print, and write your answers in the blanks provided.

LOCK BUTTON PLATE QUIZ

1. What style of title block is used on this print?

1. _____

2. What size drawing was used?

2. _____

3. List the decimal dimensions shown on the print that have unspecified tolerances.

3. _____

4. What is the part number?

4. _____

5. Name the kind of material used to make the part.

5. _____

6. What limits are shown for the diameter of the two holes?

6. _____

7. What does ECO mean?

7. _____

8. To what scale was the drawing made?

8. _____

9. What radius is given for the four corners?

9. _____

10. What operation must be performed on the workpiece according to the note on the print?

10. _____

11. What kind of finish is required on the workpiece?

11. _____

12. What specific change resulted from revision Ⓑ ?

12. _____

13. List all the high limits shown on the print.

13. _____

14. What is the drawing number?

14. _____

15. What tolerance is used on the fractional dimensions?

15. _____

16. What specific kind of tolerance is dimension .876 called?

16. _____

17. What is the low limit of dimension .438?

17. _____

18. What tolerance is used for four place decimal dimensions?

18. _____

19. What is the low limit of the 5/8 fractional dimension?

19. _____

20. The removal of a carburizing operation is specified by what ECO number?

20. _____

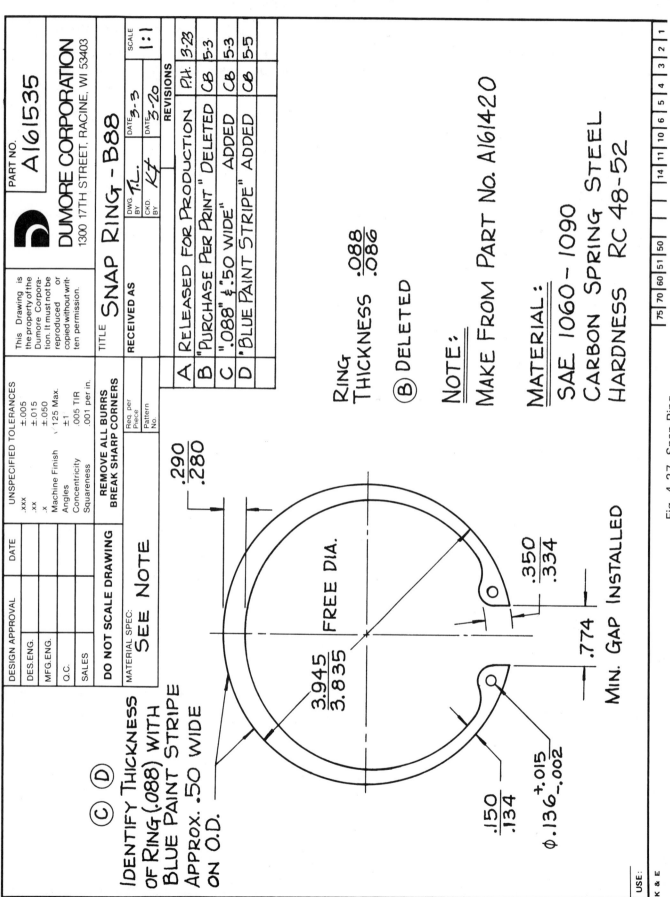

Fig. 4-27. Snap Ring.

SNAP RING QUIZ

1. What is the part number?
2. Name the kind of material used to make the part.
3. State what change was made by revision (D)
4. To what scale was the drawing made?
5. Who released revision (A) for production?
6. What style of title block is used on this print?
7. By whom was the print drawn?
8. What is the ring thickness?
9. How wide must the blue paint stripe be made?
10. What tolerance is specified for angles?
11. Who was responsible for making revision (B) ?
12. What is the name of the part?
13. Dimensions .280, 3.835, .134, and .334 are examples of what type of limits?
14. What tolerance is used for one place decimal dimensions?
15. What is the maximum value for unspecified surface finishes?
16. What tolerance is allowed on the .136 dimension?
17. What ring hardness is required?
18. Was design approval needed by design engineering?
19. Ring should be made from what existing part?
20. What are the limits of ring width opposite the gap?

1. _____
2. _____
3. _____
4. _____
5. _____
6. _____
7. _____
8. _____
9. _____
10. _____
11. _____
12. _____
13. _____
14. _____
15. _____
16. _____
17. _____
18. _____
19. _____
20. _____

Unit 5

WORKING DRAWINGS
ONE, TWO, AND THREE VIEW

After studying this unit, you will be able to:
□ Interpret a one-view drawing.
□ Interpret a two-view drawing.
□ Interpret a simple three-view drawing.

The units up to this point were designed in stages to explain various methods and skill-building techniques. Review the units that illustrate the various views, title block information, and drawing examples.

WORKING DRAWINGS

Working drawings are sometimes called "detailed drawings" or "detailed working drawings." A WORKING DRAWING provides all details necessary to properly describe the part. The information should include the correct views, dimensions, part number, tolerances, and other specifications for manufacturing. The working drawing is actually the shop print of that individual part.

Working drawings can be grouped into three general types: one-view drawings, two-view drawings, and three-view drawings (multiview). In each

type of working drawing, the specific information for manufacturing must be clearly and carefully located on the drawing.

ONE VIEW

Objects that are uniform in cross section, such as cylindrical parts and thin flat parts, usually appear on prints as one-view drawings.

A centerline through a cylindrical part indicates that the part is symmetrical. Diameters are specified by the symbol ϕ, a letter D, or an abbreviation DIA.

A thin, flat object which appears as a one-view drawing has its thickness listed in the materials specification area of the title block or it may be stated as a special note. See Figs. 5-1 and 5-2.

TWO VIEWS

Two-view drawings are commonly used to describe machined parts. These parts need not be very complex, but fulfill certain conditions:
1. That a third view would show no significant contours of the object.

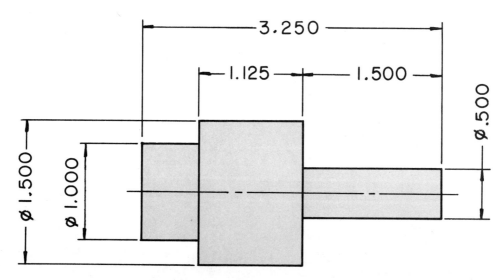

Fig. 5-1. One-view drawing—Shaft.

.033 THICK

Fig. 5-2. One-view—Gasket.

2. That two of the three views would be the same.
3. That parts are simple cylindrical, rectangular, stamped pieces, etc.
4. That no additional information, such as dimensions, would be provided by a third view.

The designer or drafter will decide which two views best represent the details of the part. The two views selected will differ depending on the part. The most common arrangement for two-view drawings is the use of top and front views as shown in Fig. 5-3. If the front and right side views were used as

alternatives, the drilled hole could be misinterpreted. Note: Top view shows drill hole by use of a circle and two centerlines. Circle in top view correctly indicates true size and shape of drilled hole.

Likewise, the front view describes the beveled ends more clearly then the top view and right-side view (not shown). As mentioned, the selection of the two views is most important to the reader for detail clarification.

In Fig. 5-4, the front view of a BEARING BLOCK shows a small circle within a larger circle. This is the best view to represent the drilled or reamed hole. Also, two angular supports are shown on either side of the large circle. The representation of both angular supports and the drill hole are found directly above on the top view. The top and front views supply the necessary details to describe the BEARING BLOCK without the addition of a third view.

THREE VIEWS

Three-view or multiview drawings are generally more complex in nature than one and two-view drawings. The three-view drawing may use any of the six principle views (see Unit 2) to illustrate the part. However, it is general practice when drawing three views to include the front view as one of the selected views.

TOP

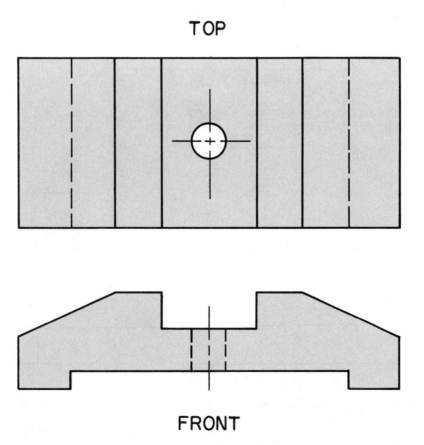

FRONT

Fig. 5-3. Two-view drawing—Guide.

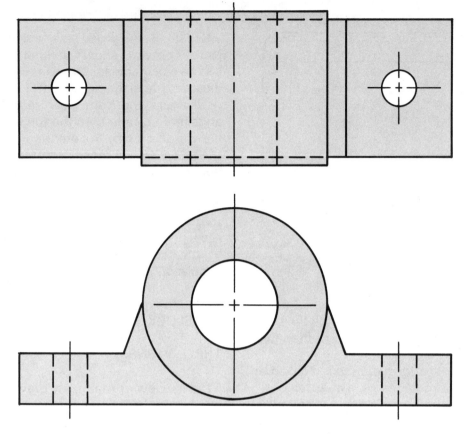

Fig. 5-4. Two-view—Bearing Block.

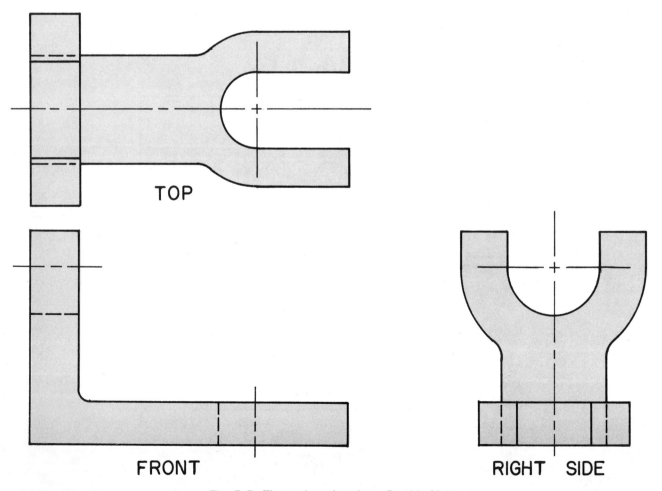

TOP

FRONT

RIGHT SIDE

Fig. 5-5. Three-view drawing—Double Yoke.

When selecting the third view (whichever view is chosen), it must be arranged in the correct position or relationship to the other two remaining views. In Fig. 5-5, the top view is correctly positioned with the front view. With the use of the right-side view, the DOUBLE YOKE becomes more clear to the reader.

The next illustration, Fig. 5-6, uses similar view arrangements: top, front, and right side. The top and front views of the FORK contain the majority of the necessary details. However, the right-side view is a most important key in describing the con-

tour of the FORK. The outer edges on the right-side view are slightly curved. This cannot be fully shown on the top or front views. The only other view which would contain the same contour information is the left-side view (not shown). The right-side view was selected because it eliminated most of the hidden lines.

Carefully examine all lines on the views before deciding on the exact shape of the part.

The dimensions were not added to Figs. 5-5 and 5-6 for line clarity reasons. The work problems that follow contain the necessary dimensions.

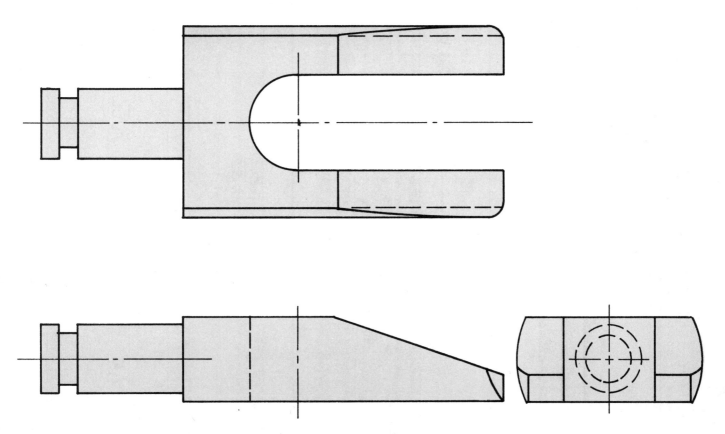

Fig. 5-6. Three-view—Fork.

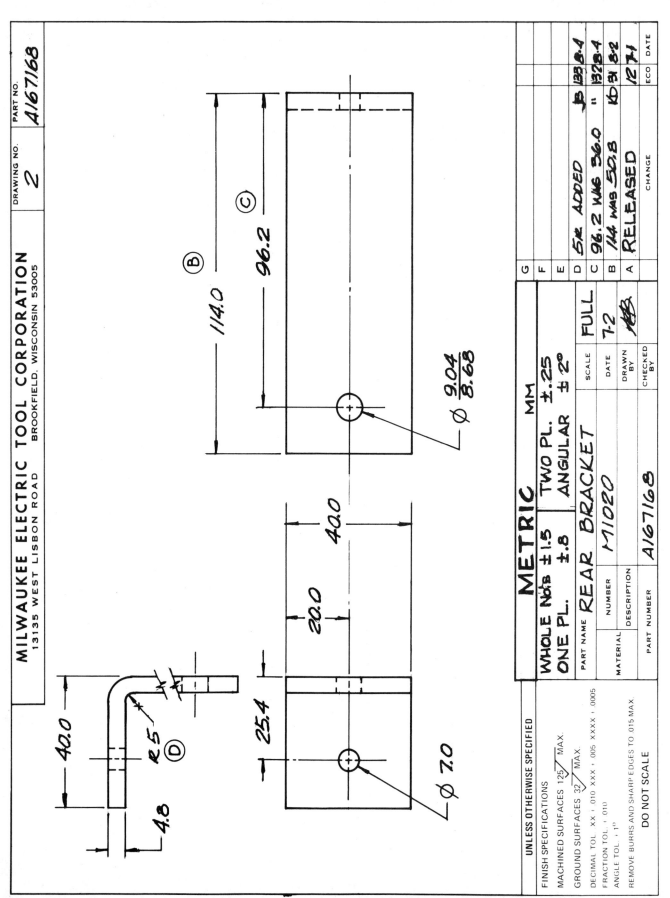

Fig. 5-7. Rear Bracket.

DIRECTIONS — QUIZ QUESTIONS

1. The industrial prints in this section will test your print reading ability.
2. Study the views, dimensions, title block, and notes in Figs. 5-7 through 5-11.
3. Read the quiz questions, refer to the print, and write your answers in the blanks provided.

REAR BRACKET QUIZ

1. How many holes does the part have?

1. _____

2. What size is the smallest hole?

2. _____

3. What system of measurement is used on the print?

3. _____

4. How long is the bracket?

4. _____

5. What is the material number?

5. _____

6. How thick is the bracket?

6. _____

7. What is the largest size to which the small diameter hole can be drilled?

7. _____

8. What was revision Ⓒ ?

8. _____

9. How far is the 9.04 diameter hole located from the end of the short leg of the bracket?

9. _____

10. What type of break line is shown in the top view?

10. _____

11. What tolerance is used on two-place metric dimensions?

11. _____

12. How long is the short leg of the bracket?

12. _____

13. What is the high limit on the 114.0 dimension?

13. _____

14. What value do the dimensions on the print represent?

14. _____

15. What does symbol ϕ represent?

15. _____

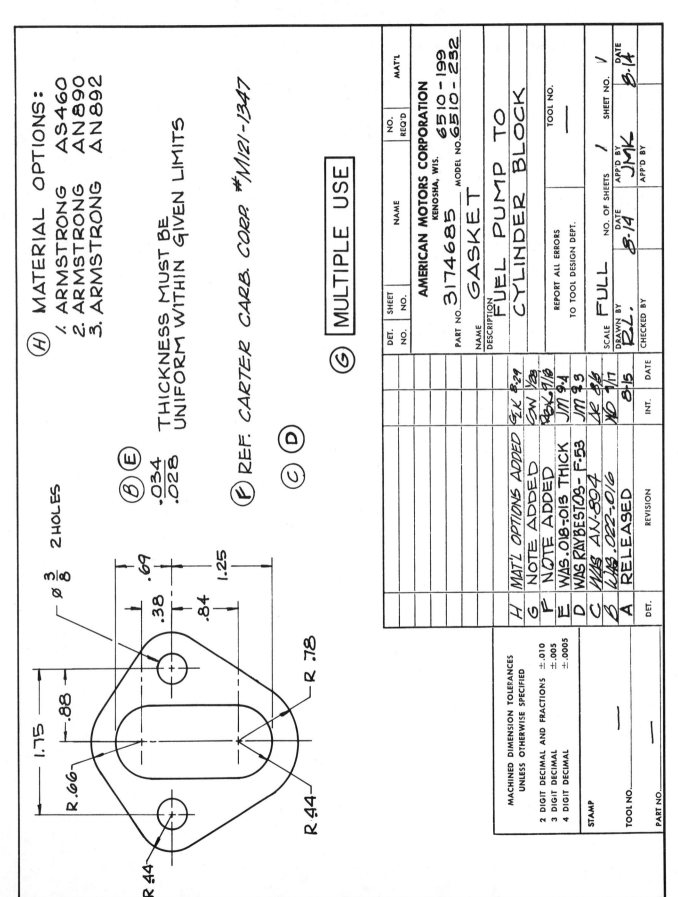

Fig. 5-8. Gasket.

Ⓗ MATERIAL OPTIONS:
1. ARMSTRONG AS460
2. ARMSTRONG AN890
3. ARMSTRONG AN892

Ⓑ Ⓔ $\frac{.034}{.028}$ THICKNESS MUST BE
UNIFORM WITHIN GIVEN LIMITS

Ⓕ REF. CARTER CARB. CORP. #M121-1347

Ⓒ Ⓓ

Ⓖ | MULTIPLE USE |

Ø $\frac{3}{8}$ 2 HOLES

.69
1.25
.38
.84
1.75
.88
R.66
R.78
R.44
R.44

MACHINED DIMENSION TOLERANCES
UNLESS OTHERWISE SPECIFIED

2 DIGIT DECIMAL AND FRACTIONS ±.010
3 DIGIT DECIMAL ±.005
4 DIGIT DECIMAL ±.0005

STAMP

TOOL NO. _____

PART NO. _____

DET. NO.	SHEET NO.	NAME	NO. REQ'D	MAT'L

AMERICAN MOTORS CORPORATION
KENOSHA, WIS.

PART NO. 3174685   MODEL NO. 6510-232
6510-199

NAME GASKET

DESCRIPTION FUEL PUMP TO
CYLINDER BLOCK

TOOL NO.
——

REPORT ALL ERRORS
TO TOOL DESIGN DEPT.

SCALE FULL	NO. OF SHEETS 1	SHEET NO. 1
DRAWN BY R.L.	DATE 8-14	APP'D BY JMK
CHECKED BY		APP'D BY JMK	DATE 8-14

DET.	REVISION	INT.	DATE
H	MAT'L OPTIONS ADDED	G.K.	8-24
G	NOTE ADDED	GW	1/23
F	NOTE ADDED	RDL	1/18
E	WAS .018-.013 THICK	JM	9-4
D	WAS RAYBESTOS - F-53	JM	9-3
C	WAS AN-894	LK	8-6
B	WAS .022-.016	WD	1/17
A	RELEASED		8-15

56

GASKET QUIZ

1. What maximum thickness is allowed on the part?   1. _____

2. What is the full length of the slot?   2. _____

3. Determine the center distance between the two holes?   3. _____

4. Give the high limit on the hole sizes.   4. _____

5. List the radii shown on the part print.   5. _____

6. By whom was the print drawn?   6. _____

7. What note is given regarding part thickness?   7. _____

8. How far is the right hole from the vertical centerline of the part?   8. _____

9. List model numbers using the gasket.   9. _____

10. What tolerance is used on four-digit decimal dimensions?   10. _____

11. List the part description.   11. _____

12. What scale is the print?   12. _____

13. How wide is the slot?   13. _____

14. How thick was the gasket when originally manufactured?   14. _____

15. What material was used to make the gasket?   15. _____

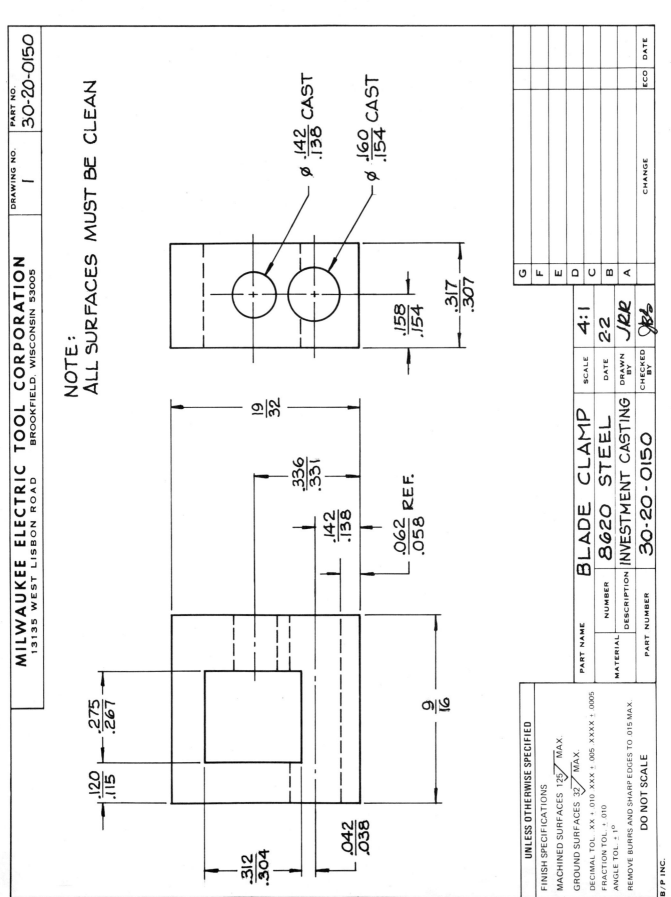

Fig. 5-9. Blade Clamp.

BLADE CLAMP QUIZ

1. What size is the rectangular hole in the part?

2. How long is the part?

3. How high is the part?

4. What tolerance is used on fractional dimensions?

5. Which hole does not go through the entire part?

6. What is the low limit of the diameter of the large cast hole?

7. What is the high limit of the diameter of the small cast hole?

8. What is the center distance between the two cast diameter holes?

9. What scale is the print?

10. Give a description of the material used to make the part?

11. What maximum part thickness is allowed?

12. How far does the rectangular hole go into the large cast diameter hole?

13. What is the material number of the part?

14. What type of line shows the rectangular hole in the side view?

15. What would the wall thickness be between the two cast holes using the mean dimension for each cast hole?

16. Is the .142 hole in the center of the rectangular hole?

17. What is the mean dimension on the long side of the rectangular hole?

18. Are the cast holes in the center of the part?

19. What is the high limit of the 9/16 dimension?

20. Is the shape of the part square or rectangular?

1. _____

2. _____

3. _____

4. _____

5. _____

6. _____

7. _____

8. _____

9. _____

10. _____

11. _____

12. _____

13. _____

14. _____

15. _____

16. _____

17. _____

18. _____

19. _____

20. _____

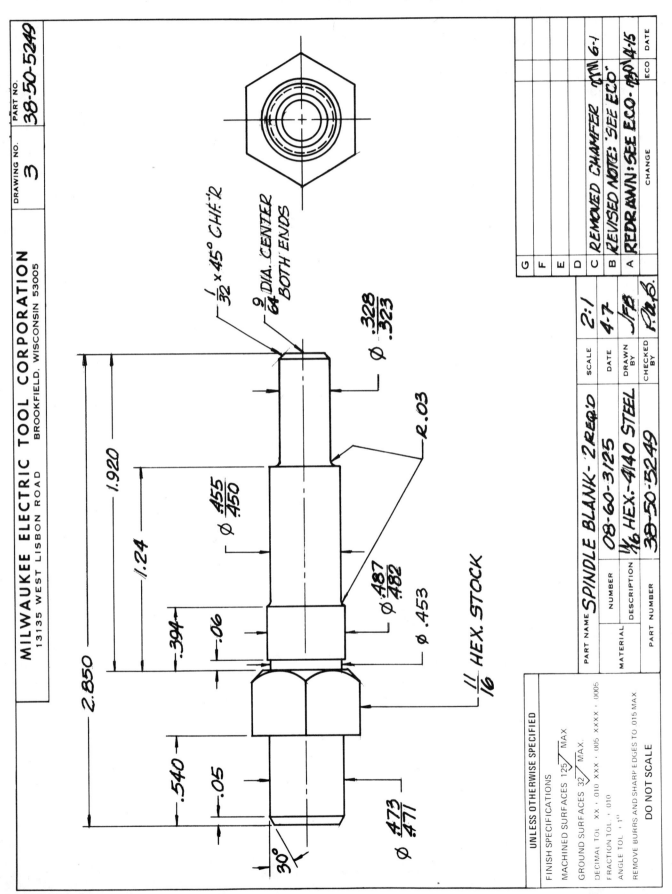

Fig. 5-10. Spindle Blank.

60

SPINDLE BLANK QUIZ

1. What size is the smallest diameter?

1. _____

2. What is the maximum overall length allowed for the workpiece?

2. _____

3. How long is the .455/.450 diameter?

3. _____

4. How long is the .487/.482 diameter?

4. _____

5. How many shoulders have a radius?

5. _____

6. What scale is the print?

6. _____

7. What tolerance is used on the .453 diameter?

7. _____

8. What length is the 30° chamfer?

8. _____

9. How long is the hex portion of the workpiece?

9. _____

10. What specific kind of material is used to make the workpiece?

10. _____

11. What finish is required on all machined surfaces?

11. _____

12. How long is the .328/.323 diameter?

12. _____

13. What is the distance from the shoulder formed by the .473 diameter and the hex to the shoulder formed by the .328 diameter and the .455 diameter?

13. _____

14. By whom was the print drawn?

14. _____

15. What is the distance from the flat of the hex to the surface of the .473 diameter?

15. _____

16. List all the low limits on the specified decimal dimensions.

16. _____

17. How many degrees is the 1/32 chamfer?

17. _____

18. Which decimal dimension has the smallest tolerance?

18. _____

19. What is the low limit for the .394 dimension?

19. _____

20. What size hex stock is used to machine the workpiece?

20. _____

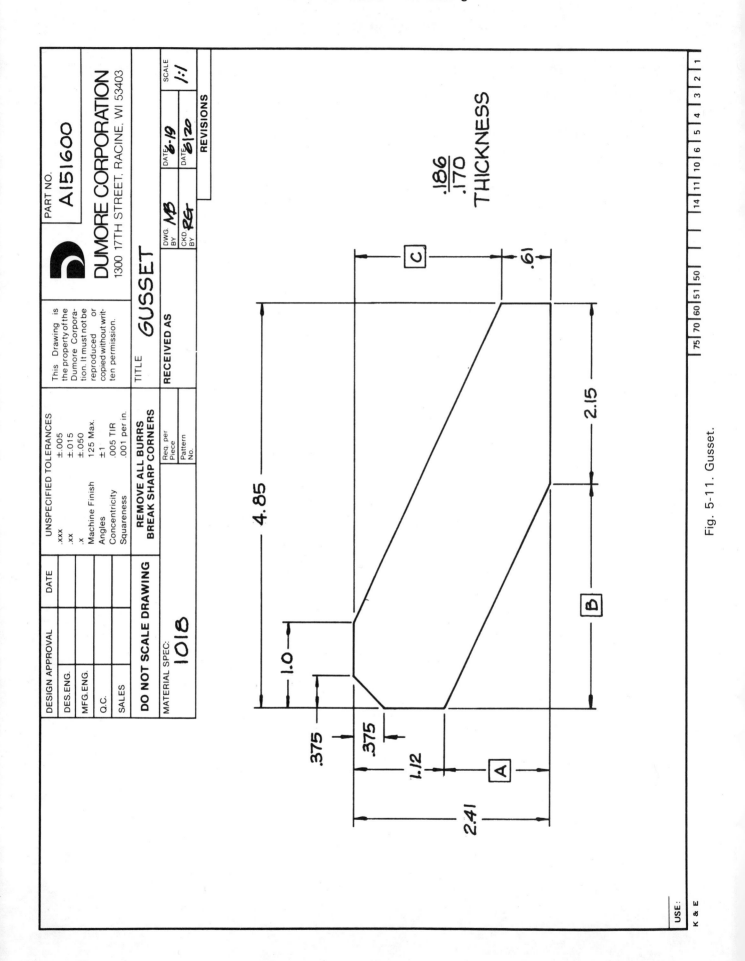

Fig. 5-11. Gusset.

GUSSET QUIZ

1. How thick is the part?

2. What material specifications are given?

3. What is the minimum overall length allowed for the part?

4. How long is dimension (A) ?

5. What is the high limit on the 2.15 dimension?

6. What is the high limit on the .375 dimension?

7. How long is dimension (B) ?

8. By whom was the print drawn?

9. Are there any revisions made on the print?

10. Who checked the print?

11. What tolerance is used on one-place decimal dimensions?

12. How long is dimension (C) ?

13. What scale is the print?

14. What directions are given on the print in regard to sharp corners?

15. What minimum part thickness is allowed?

1. _____

2. _____

3. _____

4. _____

5. _____

6. _____

7. _____

8. _____

9. _____

10. _____

11. _____

12. _____

13. _____

14. _____

15. _____

Unit 6

HOLES

After studying this unit, you will be able to:
☐ Define the terms drilling, reaming, countersinking, counterboring, spotfacing, and boring.
☐ Identify and/or list the location and size of drilled, reamed, countersunk, counterbored, spotfaced, and bored holes.

Various machining operations are used to make holes, enlarge holes, finish holes, or prepare a smooth surface around holes. These operations include drilling, reaming, countersinking, counterboring, spotfacing, and boring.

DRILLED HOLE

Drilling a hole into or through solid material is a machining operation performed by a cutting tool called a "drill."

DRILLED HOLES generally are classified into two groups:
1. THROUGH HOLES which pass completely through the part or material.
2. BLIND HOLES which are drilled partially into the object at specified depths.

On through holes, hidden lines (dashed) are drawn, touching the top and bottom surfaces of the front view. On blind holes, the drill does not pass completely through the workpiece. The drill will produce a conical point area at the bottom of the hole. Note where the 7/8 in. depth is measured in Fig. 6-1.

The drilled hole on the print contains the necessary information regarding size or diameter of drill, depth of hole, and quantity of holes to be

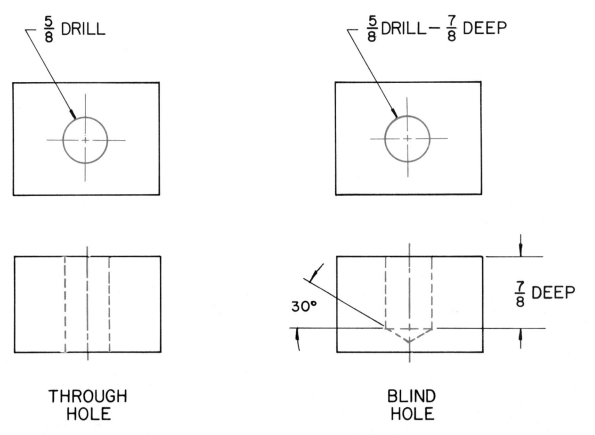

THROUGH HOLE

BLIND HOLE

Fig. 6-1. Examples of two types of drilled holes.

64

drilled. In specific situations, the drill holes may be dimensioned by equal spacing, or through angular dimensioning for circular pieces. See Fig. 6-2.

REAMED HOLE

REAMING is the operation that finishes a hole to a specific size and required finish. Reaming usually is preceded by drilling and/or boring. A small amount of material is removed by a multi-fluted tool. Material allowance for reaming usually ranges from 1/64 in. for hole sizes under 1 in. in diameter to 1/32 in. or more for holes greater than 1 in. in diameter.

The reamed hole is the same in appearance to the circle of a drilled hole. It would be very difficult and confusing to draw two circles very close to one another for the drilled hole and one for the reamed hole. The drafter simply notes the size of the drill and size of the ream on one circle. See Fig. 6-3.

COUNTERSUNK HOLE

COUNTERSINKING is the operation of enlarging the end of a hole conically to recess the head of a flat head screw. See Fig. 6-4. Drill hole size,

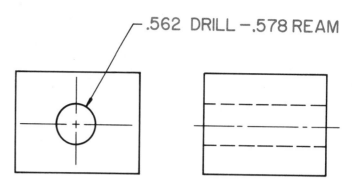

Fig. 6-3. How a reamed hole may be dimensioned.

countersink angle, and countersink diameter are specified (noted) on the circular presentation of a countersunk hole.

COUNTERBORED HOLE

COUNTERBORING is the operation of cylindrically enlarging the diameter of a previously formed hole to a specific depth. Its purpose is to provide a recess for fastener heads or a seat for bearings. The size of hole, diameter of counterbore, and depth are included in the note on the circular representation of the counterbore. See Fig. 6-5.

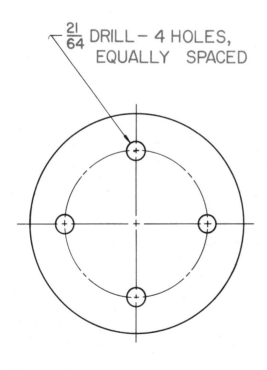

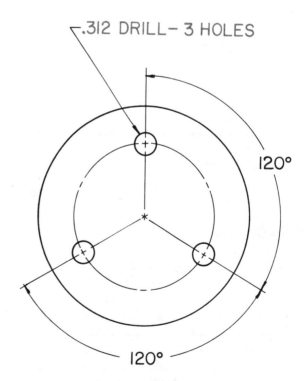

Fig. 6-2. Two ways drilled holes may be dimensioned on a print.

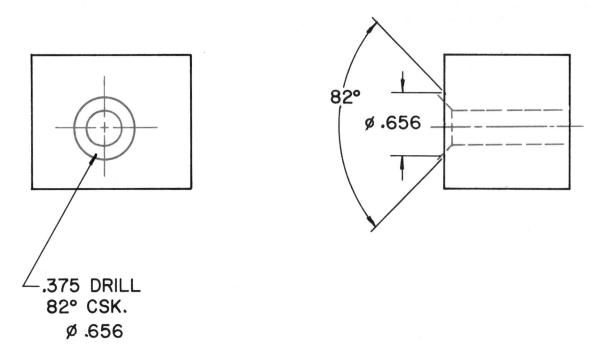

.375 DRILL
82° CSK.
Ø .656

Fig. 6-4. Representation of a countersunk hole.

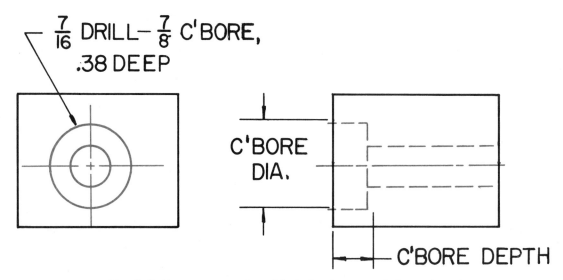

$\frac{7}{16}$ DRILL – $\frac{7}{8}$ C'BORE, .38 DEEP

C'BORE DIA.

C'BORE DEPTH

Fig. 6-5. How a counterbored hole is shown and dimensioned.

SPOTFACE HOLE

SPOTFACING is the operation of providing a smooth flat surface around a hole. This operation usually is performed on a rough surface of a casting to accommodate the seating of a washer or bolt head. Spotfacing and counterboring are machined similarly, with the exception of the depth. Spotfacing usually is shallower, and approximately 1/16 in. deep. The size of hole, spotface diameter, and depth are included in the note on the circular representation of the spotface. See Fig. 6-6.

BORED HOLE

BORING is the operation of enlarging a hole to a close tolerance and fine finish. It produces a straight and round hole more accurately than by drilling. A boring bar and single point tool are usually used to produce holes which can range in infinite diameter sizes. The representation of a bored hole appears the same as for a drilled hole and a reamed hole. The size of the hole is noted on the circular representation of the bored hole, as shown in the top view of the part in Fig. 6-7.

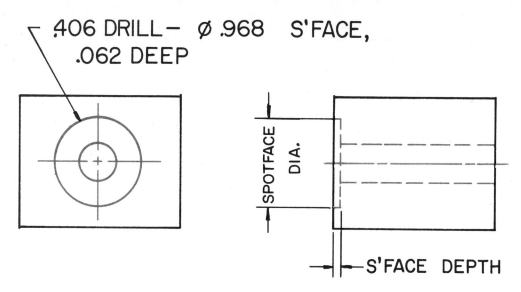

.406 DRILL— Ø .968 S'FACE,
.062 DEEP

SPOTFACE DIA.

S'FACE DEPTH

Fig. 6-6. Spotface hole dimensioning.

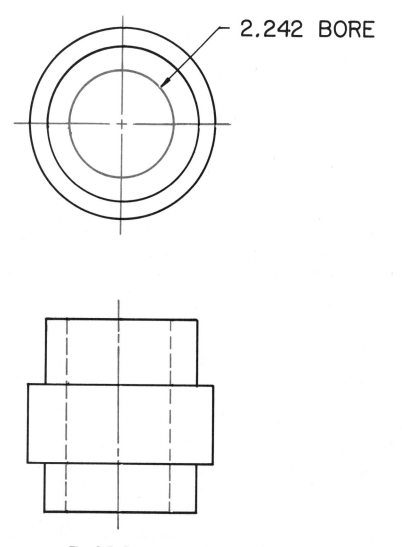

2.242 BORE

Fig. 6-7. Representation of a bored hole.

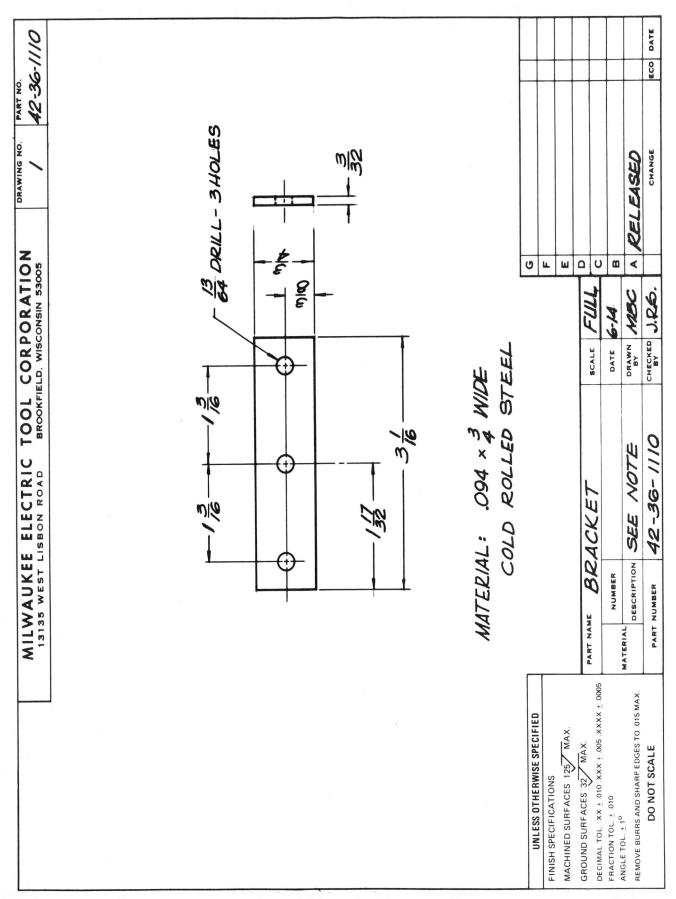

Fig. 6-8. Bracket.

DIRECTIONS—QUIZ QUESTIONS

1. The industrial prints in this section will test your print reading ability.
2. Study the views, dimensions, title block, and notes in Figs. 6-8 through 6-11.
3. Read the quiz questions, refer to the print, and write your answers in the blanks provided.

BRACKET QUIZ

1. What is the overall length of the workpiece?

2. What size holes are drilled through the workpiece?

3. What is the scale of the blueprint?

4. What is the part name?

5. What kind of material is specified on the blueprint?

6. Is the width of this workpiece machined?

7. What is the distance from the left end of the workpiece to the centerline of the left hole?

8. What is the fractional thickness of the workpiece?

9. What is the decimal thickness of the workpiece?

10. What is the distance between the two extreme holes?

11. What tolerance is used on fractional dimensions?

12. What is the part number of the workpiece?

13. What is the high limit on the overall length of the workpiece?

14. What is the smallest size the holes can be drilled?

15. Are there any revisions on the blueprint?

1. _____

2. _____

3. _____

4. _____

5. _____

6. _____

7. _____

8. _____

9. _____

10. _____

11. _____

12. _____

13. _____

14. _____

15. _____

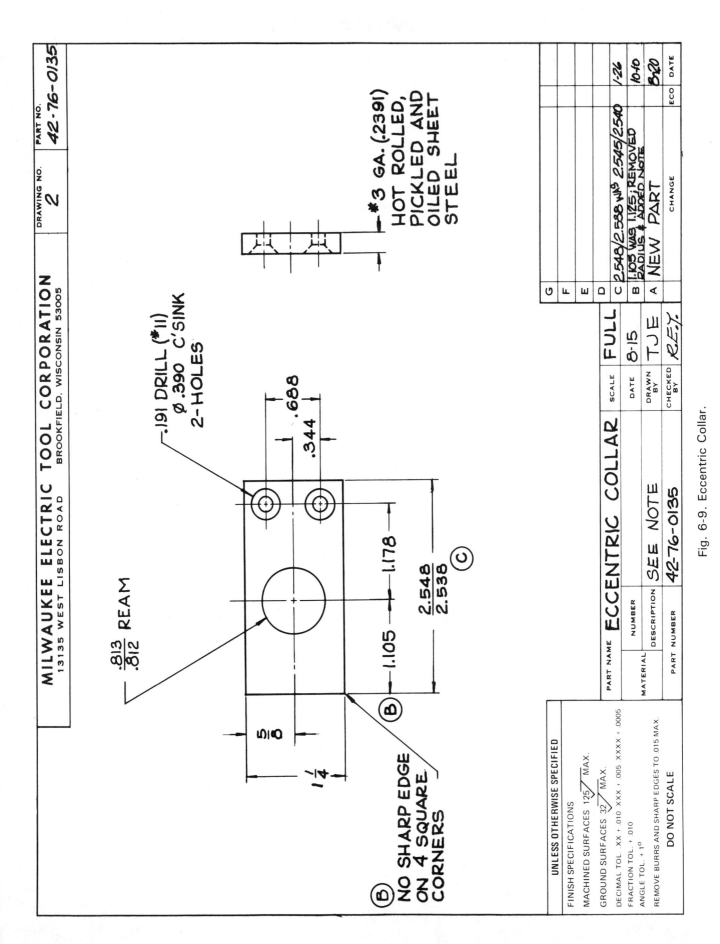

MILWAUKEE ELECTRIC TOOL CORPORATION
13135 WEST LISBON ROAD   BROOKFIELD, WISCONSIN 53005

DRAWING NO. 2

PART NO. 42-76-0135

*3 GA. (.2391) HOT ROLLED, PICKLED AND OILED SHEET STEEL

.191 DRILL (#11)
Ø .390 C'SINK
2-HOLES

.813 / .812 REAM

.688
.344

1.178

2.548 / 2.538

C

1.105

B

5/8

1 1/4

B NO SHARP EDGE ON 4 SQUARE CORNERS

	CHANGE	ECO	DATE
G			
F			
E			
D			
C	2.548/2.538 WAS 2.545/2.540		1-26
B	1.105 WAS 1.125; REMOVED RADIUS & ADDED NOTE		10-10
A	NEW PART		8-20

PART NAME ECCENTRIC COLLAR

SCALE FULL

DATE 8-15

DRAWN BY TJE

CHECKED BY R.E.J.

MATERIAL	NUMBER	DESCRIPTION	SEE NOTE
PART NUMBER	42-76-0135		

UNLESS OTHERWISE SPECIFIED

FINISH SPECIFICATIONS

MACHINED SURFACES 125√ MAX.

GROUND SURFACES 32√ MAX.

DECIMAL TOL. .XX ± .010 .XXX ± .005 .XXXX ± .0005

FRACTION TOL. ± .010

ANGLE TOL. ± 1°

REMOVE BURRS AND SHARP EDGES TO .015 MAX

DO NOT SCALE

Fig. 6-9. Eccentric Collar.

ECCENTRIC COLLAR QUIZ

1. How far apart are the two small holes from the vertical centerline of the large hole?

2. What is the distance between the two small holes?

3. What is the overall length of the workpiece?

4. What is the low limit of the length dimension?

5. What size is the large hole?

6. How thick is the workpiece?

7. How wide is the workpiece?

8. How far is the vertical centerline of the large hole from the left end of the workpiece?

9. What was revision C?

10. How is the large hole made?

11. What is the farthest the two small holes can be spaced from each other?

12. How many corners have a radius?

13. How much wall thickness is there from the edge of the small holes to the right side of the workpiece?

14. What is the distance from horizontal centerline of workpiece to centerline of upper small hole?

15. What scale is the print?

16. What tolerance is allowed on three decimal dimensions?

17. What kind of material is the workpiece?

18. How much tolerance is allowed on the large hole?

19. To what size are the two holes countersunk?

20. What is the high limit on the width of the workpiece?

1. _____

2. _____

3. _____

4. _____

5. _____

6. _____

7. _____

8. _____

9. _____

10. _____

11. _____

12. _____

13. _____

14. _____

15. _____

16. _____

17. _____

18. _____

19. _____

20. _____

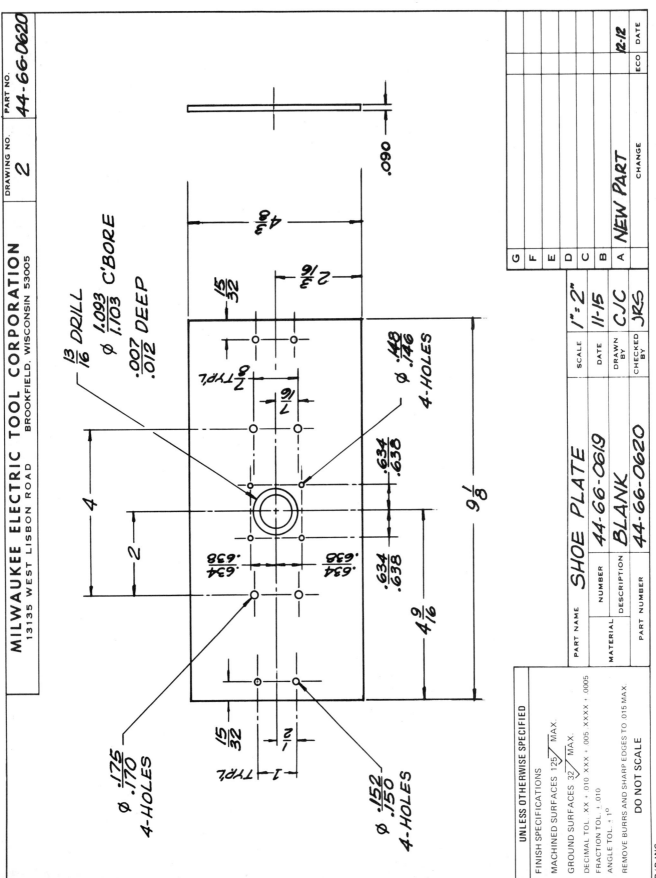

Fig. 6-10. Shoe Plate.

SHOE PLATE QUIZ

1. What is the overall length of the workpiece?

2. What is the high limit on the overall length?

3. How thick is the workpiece?

4. What size is the large hole?

5. What is the high limit on size of the large hole?

6. What is the part number?

7. How far apart are the .152/.150 holes from the horizontal centerline of the workpiece?

8. How many .152/.150 holes are there?

9. How far apart are the two .152/.150 holes on the left from the two .152/.150 holes on the right?

10. How far apart are the two .175/.170 holes on the right from the vertical centerline of the workpiece?

11. How far apart are the .175/.170 holes from the horizontal centerline of the workpiece?

12. How far apart are the two .152/.150 holes on the left from the two .175/.170 holes on the left?

13. How far are the two top .148/.146 holes from the top edge of the workpiece?

14. How far apart are the two .148/.146 holes on the right from the vertical centerline of the workpiece?

15. How far apart are the two .175/.170 holes on the left from the two .175/.170 holes on the right?

16. How deep is the counterbore?

17. How much tolerance is given on the counterbore diameter?

18. What is the total number of small drilled holes?

19. What is the smallest diameter size you can make the counterbore?

20. What is the wall thickness from the .152/.150 holes to the sides of the workpiece?

1. _____

2. _____

3. _____

4. _____

5. _____

6. _____

7. _____

8. _____

9. _____

10. _____

11. _____

12. _____

13. _____

14. _____

15. _____

16. _____

17. _____

18. _____

19. _____

20. _____

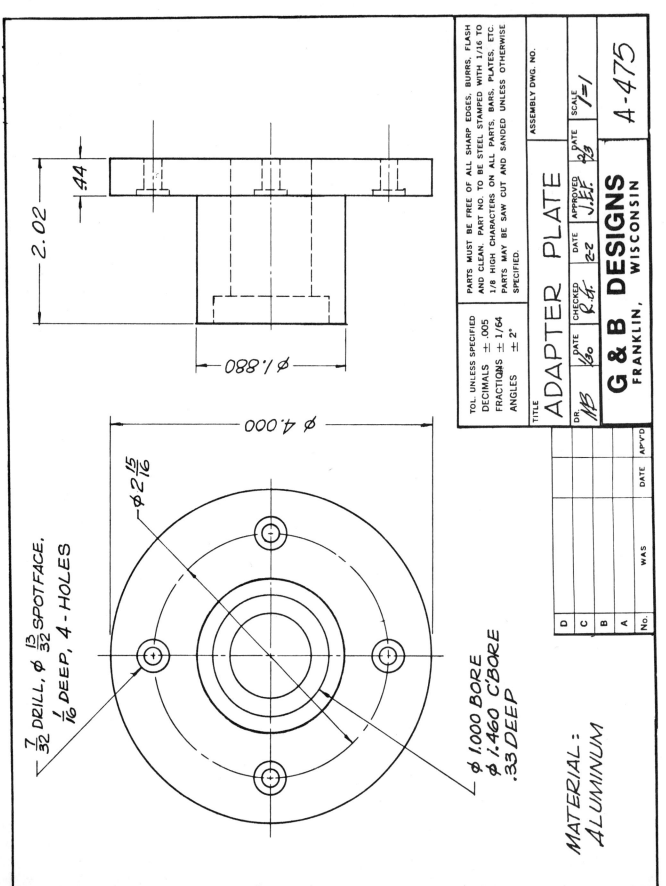

Fig. 6-11. Adapter Plate.

2.02

.44

ϕ 1.880

ϕ 4.000

ϕ 2 $\frac{15}{16}$

$\frac{7}{32}$ DRILL, ϕ $\frac{13}{32}$ SPOTFACE,
$\frac{1}{16}$ DEEP, 4 - HOLES

ϕ 1.000 BORE
ϕ 1.460 C'BORE
.33 DEEP

MATERIAL:
ALUMINUM

PARTS MUST BE FREE OF ALL SHARP EDGES, BURRS, FLASH
AND CLEAN. PART NO. TO BE STEEL STAMPED WITH 1/16 TO
1/8 HIGH CHARACTERS ON ALL PARTS, BARS, PLATES, ETC.
PARTS MAY BE SAW CUT AND SANDED UNLESS OTHERWISE
SPECIFIED.

TOL. UNLESS SPECIFIED
DECIMALS ± .005
FRACTIONS ± 1/64
ANGLES ± 2°

TITLE
ADAPTER PLATE

ASSEMBLY DWG. NO.

	DATE		DATE		DATE
DR. MB	Go	CHECKED R.G.	2-2	APPROVED J.E.F.	2 3

G & B DESIGNS
FRANKLIN, WISCONSIN

SCALE 1=1

A-475

D			
C			
B			
A			
No.	WAS	DATE	APVD

ADAPTER PLATE QUIZ

1. What is the total length of the workpiece?

2. From what material is the part made?

3. How deep is the counterbore?

4. What is the high limit on the large outside diameter?

5. How long is the 1.880 diameter?

6. What is the scale of the blueprint?

7. What tolerance is allowed for fractions?

8. What is the smallest size the bore can be machined?

9. How many holes are spotfaced?

10. What diameter size are the spotfaced holes?

11. How thick is the flange?

12. What tolerance is allowed for angles?

13. Are the spotfaced holes equally spaced?

14. What is the largest size to which the bore can be machined?

15. On what bolt circle are the 7/32 diameter holes located?

16. Is the centerline of the spotfaced holes in the center of the flange?

17. How many degrees apart are the spotfaced holes?

18. What is the low limit angular dimension between two adjacent spotfaced holes?

19. If the counterbore were machined to a depth of .325, what would be the length of the bored hole?

20. What wall thickness is there between the counterbore and the 1.880 diameter?

1. _____

2. _____

3. _____

4. _____

5. _____

6. _____

7. _____

8. _____

9. _____

10. _____

11. _____

12. _____

13. _____

14. _____

15. _____

16. _____

17. _____

18. _____

19. _____

20. _____

Unit 7

THREADS

After studying this unit, you will be able to:
☐ Describe similarities and differences between several 60° thread forms.
☐ Recognize a thread designation on an industrial print.
☐ Interpret the meaning of thread specifications on an industrial print.

Threads and screw threads appear on many machine shop blueprints. This unit describes thread uses, thread series, thread forms, thread specifications, thread representations, ISO metric threads, and thread designation.

THREAD USES

Threads are used in several situations:
1. Fastening or joining of two or more parts.
2. Conveying power (most commonly found on machine tools in the form of a lead screw).

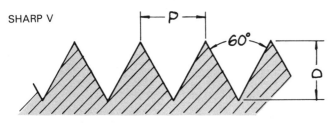

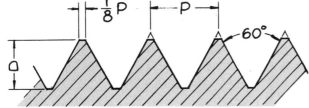

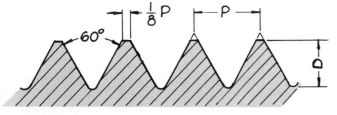

Fig. 7-1. Thread forms.

3. Providing motion or travel (as in the case of measuring tools).

Threads can be either external (outside), as in bolts and screws, or internal (inside), as in hexagonal (six-sided) nuts and square nuts.

THREAD SERIES

Threads are available in various series. These series include coarse threads, fine threads, extra-fine threads, and constant pitch threads.

COARSE THREADS

Coarse threads are used for general applications. They are specified as UNC (Unified National Coarse).

FINE THREADS

Fine threads are used in situations calling for high strength and vibration resistance. They are specified as UNF (Unified National Fine).

EXTRA-FINE THREADS

Extra-fine threads are used for special applications, such as thin wall tubing, nuts, or couplings. Extra-fine threads are specified as UNEF (Unified National Extra-Fine).

CONSTANT PITCH THREADS

Constant pitch threads are also used for special applications. They are specified as UN, with the numbers 4, 6, 8, 12, 16, 20, 28, or 32 representing the number of threads per inch.

THREAD FORMS

Two of the most common forms (shapes) of threads used are the UN (Unified National) and the N (American National) threads, Fig. 7-1. These threads are similar in form, looking like a "V" and having an included angle of 60° between the sides

76

of the thread. The only differences between the threads are the shape of the tops (crest) and the shape of the bottom (root) of the threads.

Another form of thread not widely used, but similar to UN and N threads, is the Sharp V thread, also shown in Fig. 7-1.

THREAD SPECIFICATIONS

Various specifications are given, Fig. 7-2, when representing screw threads on a drawing:
1. Outside diameter of thread (major diameter).
2. Number of threads per inch.
3. Thread form (shape or profile of thread).
    Example: U (Unified) N (National)
   Thread series (groups of diameter-pitch combinations differing from each others by number of threads per inch applied to specific diameters).
    Example: C-Coarse series, F-Fine series, EF-Extra-Fine series, 4N-4 Pitch series, etc.
4. Class fit—Fits range from a loose fit of No. 1 to a tight fit of No. 3. Fit pertains to how difficult or how easy a thread assembles with a mating thread. Specifications are in a reference manual such as Machinery's Handbook.

5. External or internal thread—Letter A is used to represent an external thread. Letter B represents an internal thread.
6. Right or left-hand thread—Unless designated, threads are right handed. When threads are left-handed, the letters LH are added at the end of the thread designation.

THREAD REPRESENTATIONS

Screw threads are represented on a drawing by different methods. See Fig. 7-3.

TAPPED HOLES

A threaded hole is represented on a print as shown in Fig. 7-4. The depth of thread (7/8 in.) is indicated in the sectioned view.

ISO METRIC THREADS

Metric screw thread standards are established by the International Organization for Standardization (ISO) in 1949.

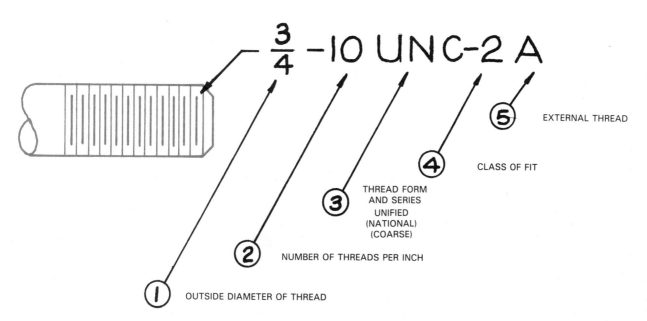

Fig. 7-2. Thread specifications.

SCHEMATIC

SIMPLIFIED

Fig. 7-3. Thread representation.

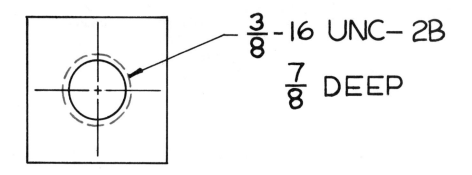

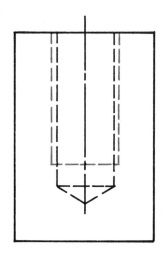

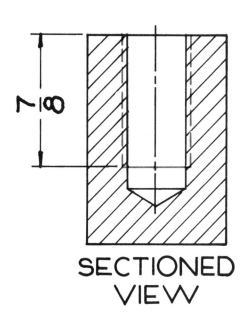

SECTIONED
VIEW

Fig. 7-4. Threaded hole.

METRIC THREAD SERIES

There are three series of metric threads:
1. ISO metric coarse pitch.
2. ISO metric fine pitch.
3. ISO metric constant pitch.

Although ISO metric and Unified share the same thread form of 60 degrees, they are not interchangable due to differences in diameters and pitches.

METRIC THREAD USE

ISO metric coarse pitch threads are in common use on fasteners. ISO metric fine pitch threads are chiefly used on precision measuring tools and instruments. The ISO metric constant pitch series is found on machine parts and all spark plugs.

THREAD DESIGNATION

Designations for all ISO metric threads begin with the capital letter "M." Next, the basic major diameter is specified in millimeters. An "X" follows; then, finally, the pitch is specified in millimeters. See Fig. 7-5.

When designating coarse series threads, the pitch is omitted. For example: A 16 mm coarse series thread with a pitch of 2 mm would be simply designated M16. The same diameter thread in the fine series would be designated M16 x 1.5.

Threads

M16

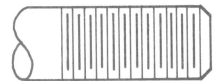

METRIC DIA. (mm)   PITCH (mm)
M16 x 1.5

METRIC DIA. (mm)
M16 x 1.5

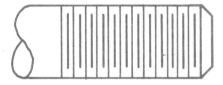

Fig. 7-5. Metric thread designation.

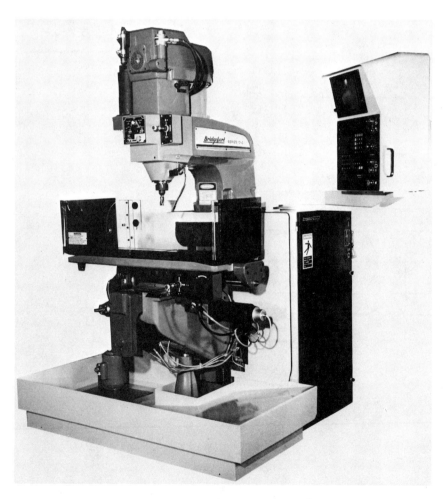

This vertical milling machine has two levels of computer-assisted numerical control programming. The first workpiece is machined by manual operation. Next, machining coordinates are entered and recorded in memory. Then, succeeding parts are machined automatically. (Bridgeport Machines, A Division of Textron, Inc.)

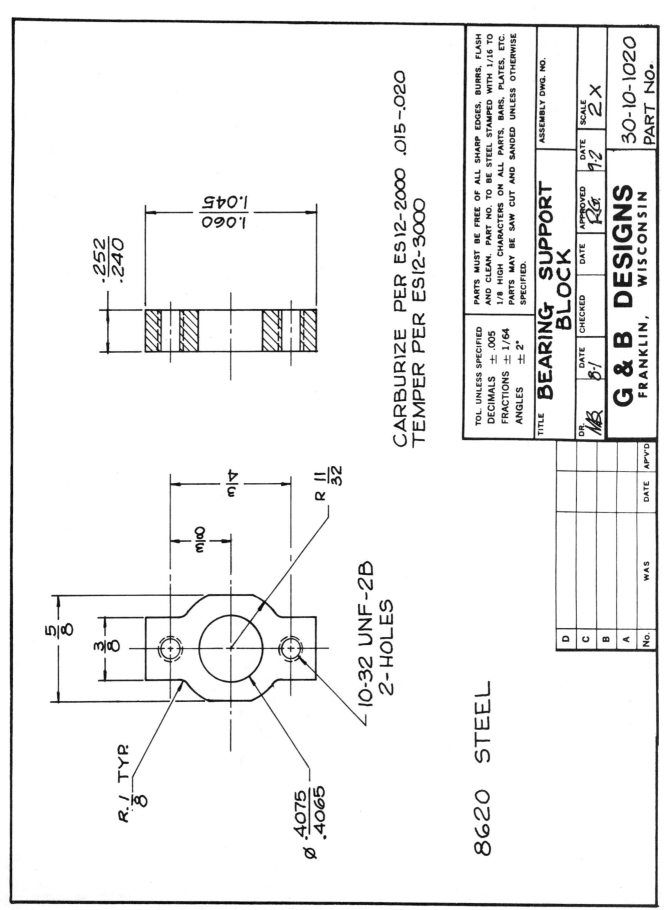

CARBURIZE PER ES12-2000 .015-.020
TEMPER PER ES12-3000

8620 STEEL

R.1 TYP.
8

Ø .4075
.4065

10-32 UNF-2B
2-HOLES

R 11
32

3
8

5
8

3
8

4
3

.252
.240

1.060
1.045

TOL. UNLESS SPECIFIED					ASSEMBLY DWG. NO.
DECIMALS ± .005 FRACTIONS ± 1/64 ANGLES ± 2°	PARTS MUST BE FREE OF ALL SHARP EDGES, BURRS, FLASH AND CLEAN. PART NO. TO BE STEEL STAMPED WITH 1/16 TO 1/8 HIGH CHARACTERS ON ALL PARTS, BARS, PLATES, ETC. PARTS MAY BE SAW CUT AND SANDED UNLESS OTHERWISE SPECIFIED.				

TITLE **BEARING SUPPORT BLOCK**

DR.	DATE	CHECKED	DATE	APPROVED	DATE
MB	8-1			RG	9-2

G & B DESIGNS
FRANKLIN, WISCONSIN

SCALE 2X

30-10-1020
PART NO.

D				
C				
B				
A				
No.		WAS	DATE	APV'D

Fig. 7-6. Bearing Support Block.

DIRECTIONS—QUIZ QUESTIONS

1. The industrial prints in this section will test your print reading ability.
2. Study the views, dimensions, title block, and notes in Figs. 7-6 through 7-10.
3. Read the quiz questions, refer to the print, and write your answers in the blanks provided.

BEARING SUPPORT BLOCK QUIZ

1. What size is the reamed hole?

   1. _____

2. What tolerance is given for the reamed hole?

   2. _____

3. What is the distance between the two tapped holes?

   3. _____

4. What size are the two threaded holes?

   4. _____

5. How far is the bottom threaded hole from the horizontal centerline of the reamed hole?

   5. _____

6. What is the total width of the workpiece?

   6. _____

7. What is the maximum length allowed on the workpiece?

   7. _____

8. What is the minimum length allowed on the workpiece?

   8. _____

9. What scale is the blueprint?

   9. _____

10. What material is used for the workpiece?

    10. _____

11. What is the minimum thickness of the workpiece?

    11. _____

12. What is the maximum thickness allowed for the workpiece?

    12. _____

13. What tolerance is allowed on four-place decimal dimensions?

    13. _____

14. What fit is called for on the threaded holes?

    14. _____

15. To what depth is the workpiece carburized?

    15. _____

16. Have there been any print revisions?

    16. _____

17. What radii are given on the print?

    17. _____

18. Which one is the fillet?

    18. _____

19. What is the name of the sectional view shown?

    19. _____

20. What heat treatment is performed according to ES 12-3000?

    20. _____

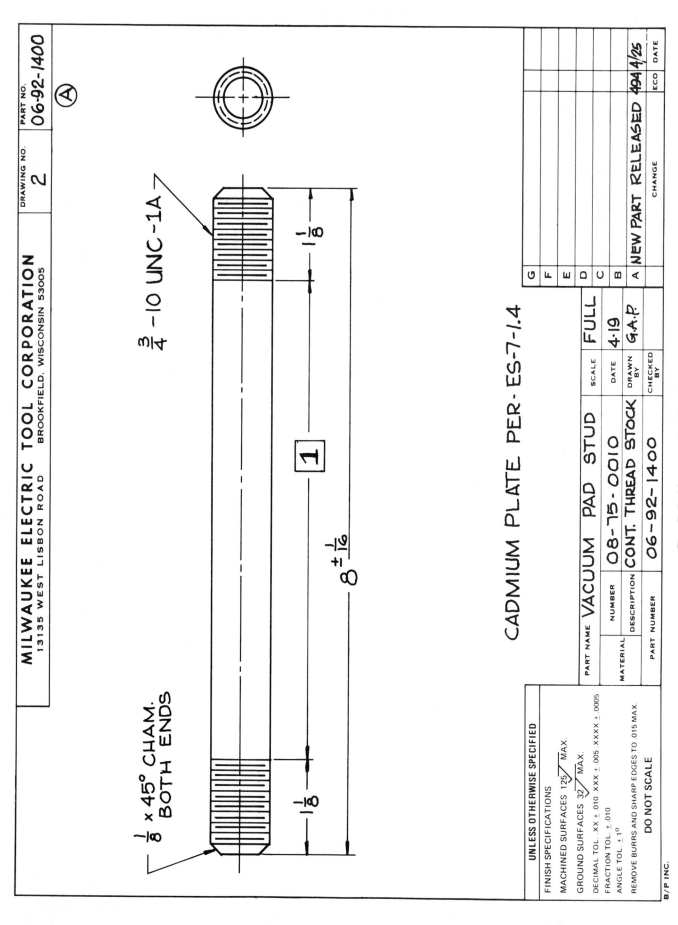

Fig. 7-7. Vacuum Pad Stud.

VACUUM PAD STUD QUIZ

1. What is the scale of the print?

1. _____

2. What two views are shown on the print?

2. _____

3. What size are the two chamfers on the workpiece?

3. _____

4. What does Ⓐ mean?

4. _____

5. What does UNC signify?

5. _____

6. How long are the threads on each end of the workpiece?

6. _____

7. What kind of thread representation is shown on the print?

7. _____

8. Who drew the print?

8. _____

9. What is the material number?

9. _____

10. How many threads per inch is the thread?

10. _____

11. What is the high limit on the overall length?

11. _____

12. What diameter is the stud?

12. _____

13. What operation is performed according to ES-7-1.4?

13. _____

14. What does ECO mean?

14. _____

15. What tolerance is given on the length of the stud?

15. _____

16. What is the value of dimension ⬚1 ?

16. _____

17. What class fit is the thread?

17. _____

18. What print change was made?

18. _____

19. What is the part number of the workpiece?

19. _____

20. What is the diameter of the thread?

20. _____

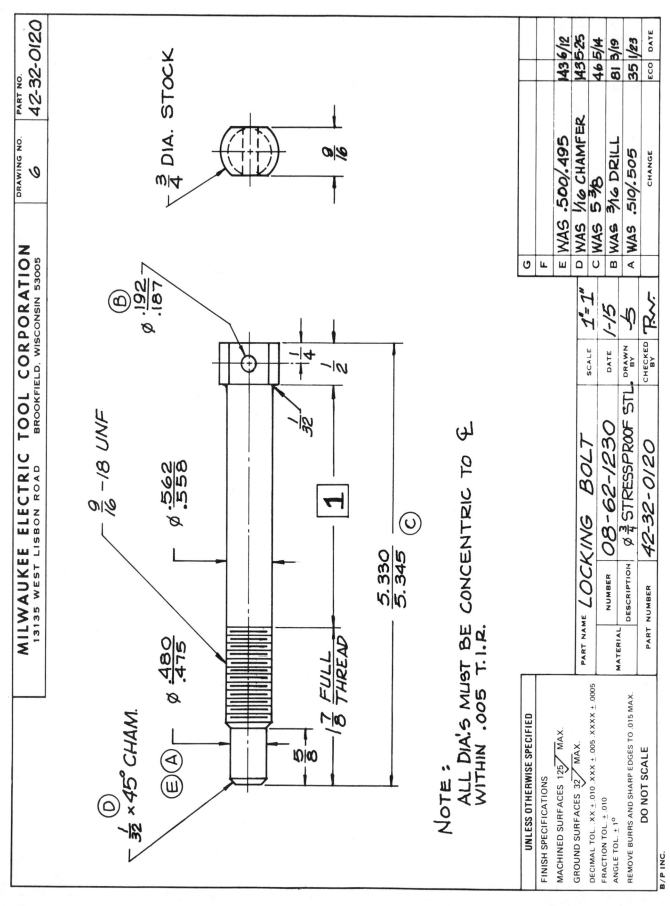

Fig. 7-8. Locking Bolt.

LOCKING BOLT QUIZ

1. What size hole is drilled in the workpiece?

2. What is the largest outside diameter?

3. How much tolerance is given on the overall length of the workpiece?

4. What print change was made at (B) ?

5. What size radius is specified on the print?

6. How long is the thread?

7. What size is the thread diameter?

8. What does UNF mean?

9. What did ECO 46 specify?

10. What material is the workpiece?

11. What is the low limit on the smallest outside diameter?

12. What is the value of dimension [1] ?

13. What is the mean dimension of the main body diameter?

14. What is the width of the locking bolt head?

15. How far is the drilled hole from the right end of the workpiece?

16. What is the scale of the print?

17. What was the original size of the smallest outside diameter?

18. How concentric must the diameters be to the centerline of the workpiece?

19. What is the low limit on the bolt head length dimension?

20. What is the tolerance for the 5/8 length dimension?

1. _____

2. _____

3. _____

4. _____

5. _____

6. _____

7. _____

8. _____

9. _____

10. _____

11. _____

12. _____

13. _____

14. _____

15. _____

16. _____

17. _____

18. _____

19. _____

20. _____

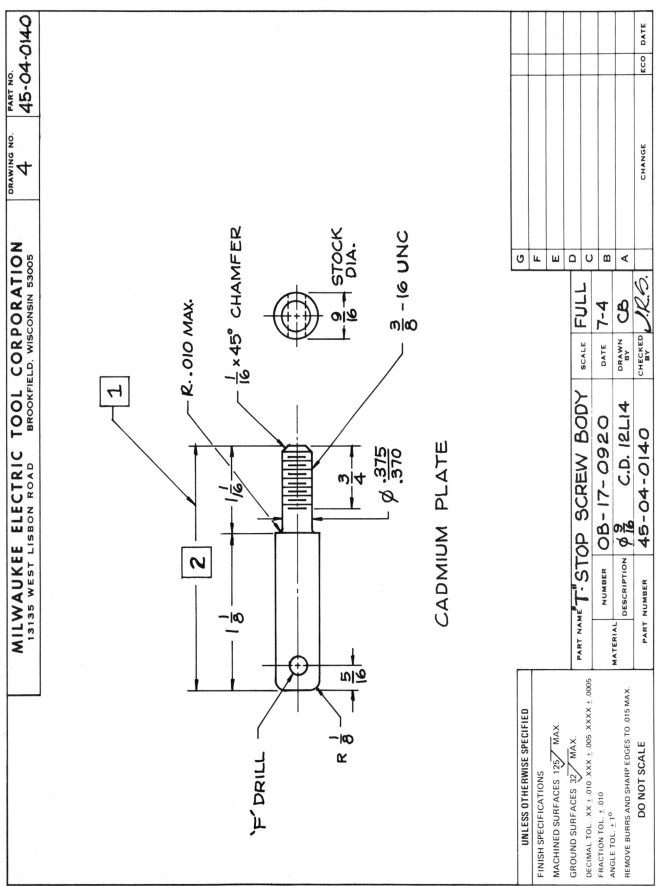

Fig. 7-9. "T"-Stop Screw Body.

"T"-STOP SCREW BODY QUIZ

1. What material number is specified for the part?

2. What size is the large diameter of the workpiece?

3. What two views are shown on the print?

4. What kind of line is line ☐1 ?

5. What size fillet is specified on the print?

6. What size hole is called for on the print?

7. Does the hole go through the workpiece?

8. How long is the thread?

9. How far is the hole located from the shoulder of the workpiece?

10. What material is the workpiece?

11. What is the part number of the print?

12. How long is the 3/8 diameter?

13. What is the value of dimension ☐2 ?

14. What is the high limit on the small diameter?

15. What is the wall thickness from the drilled hole to the left end of the workpiece?

16. What size radius is specified on the end of the workpiece?

17. What finish is specified for the workpiece?

18. What size thread is called for on the print?

19. What is the maximum length allowed on the 9/16 diameter?

20. What size chamfer is specified on the print?

1. _____

2. _____

3. _____

4. _____

5. _____

6. _____

7. _____

8. _____

9. _____

10. _____

11. _____

12. _____

13. _____

14. _____

15. _____

16. _____

17. _____

18. _____

19. _____

20. _____

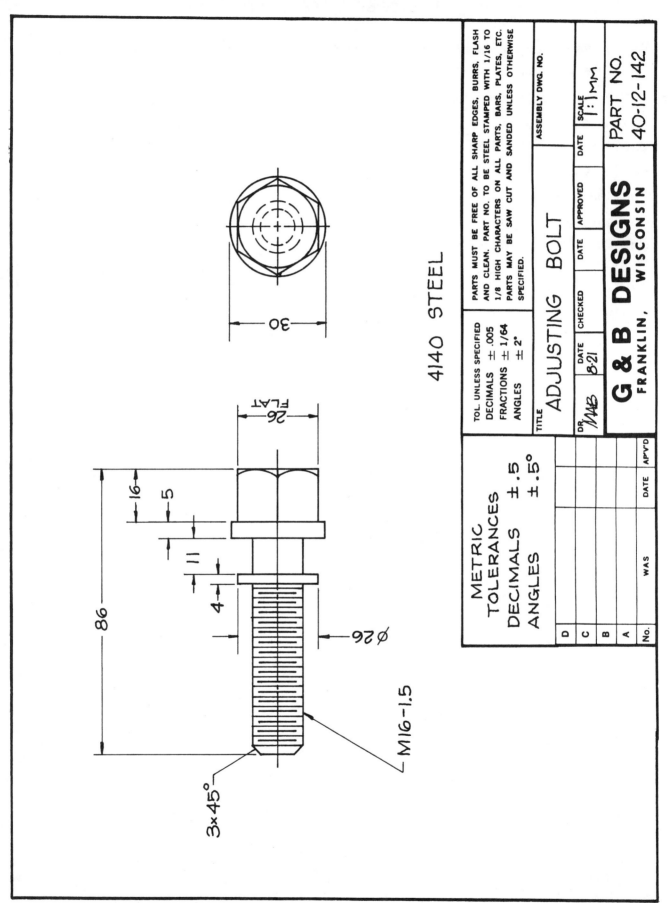

4140 STEEL

FLAT
26

16
5
11
4
86
ø26
3×45°
M16-1.5
30

METRIC
TOLERANCES
DECIMALS ±.5
ANGLES ±.5°

TOL. UNLESS SPECIFIED	PARTS MUST BE FREE OF ALL SHARP EDGES, BURRS, FLASH AND CLEAN. PART NO. TO BE STEEL STAMPED WITH 1/16 TO 1/8 HIGH CHARACTERS ON ALL PARTS, BARS, PLATES, ETC. PARTS MAY BE SAW CUT AND SANDED UNLESS OTHERWISE SPECIFIED.		
DECIMALS ± .005			
FRACTIONS ± 1/64			
ANGLES ± 2°			

TITLE
ADJUSTING BOLT

ASSEMBLY DWG. NO.

DR. MAB	DATE 8-21	CHECKED	DATE	APPROVED	DATE	SCALE 1:1 MM

G & B DESIGNS
FRANKLIN, WISCONSIN

PART NO.
40-12-142

D				
C				
B				
A				
No.	WAS	DATE	APVD	

Fig. 7-10. Adjusting Bolt.

ADJUSTING BOLT QUIZ

1. What is the high limit on the largest diameter?

   1. _____

2. What is the low limit on the total length of the workpiece?

   2. _____

3. What is the scale of the blueprint?

   3. _____

4. What tolerance is specified for angles on a metric print?

   4. _____

5. What is the length of the threaded portion of the bolt?

   5. _____

6. What is the pitch of the thread?

   6. _____

7. Is the thread coarse or fine?

   7. _____

8. What is the width of the groove on the bolt?

   8. _____

9. What is the part number of the workpiece?

   9. _____

10. What is the distance from the surface of the flat to the outside diameter of the shoulder on the bolt head?

   10. _____

11. What is the diameter of the groove?

   11. _____

12. How deep is the groove from the 26 mm diameter?

   12. _____

13. What material is the workpiece?

   13. _____

14. What is the diameter of the thread?

   14. _____

15. What is the length of the bolt head?

   15. _____

16. What tolerance is specified for the width of the groove?

   16. _____

17. Does the blueprint specify a part number?

   17. _____

18. What is the distance from the top of the bolt head to the bottom of the 26 mm diameter collar?

   18. _____

19. What thread chamfer is specified?

   19. _____

20. What is the distance from the outside of the collar diameter to the thread diameter?

   20. _____

Unit 8

CONTOURS

After studying this unit, you will be able to:
☐ Specify the size of a fillet, round, or contour.
☐ Know the difference between a fillet and a round.
☐ Calculate distances using radii dimensions.

A CONTOUR is the outline of an object, especially a curving or irregular outline. Generally, in drafting, contours are lines that represent arcs and curved surfaces.

RADII

An arc or curved surface less than an entire circle is dimensioned by its radius (R). A broken dimension line is used to give the size of the arc and to indicate that the origin of the arc lies outside the

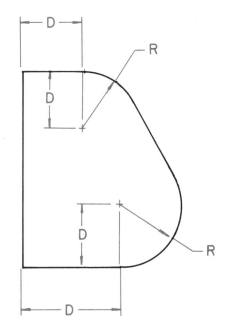

Fig. 8-2. Example of radii with defined origins.

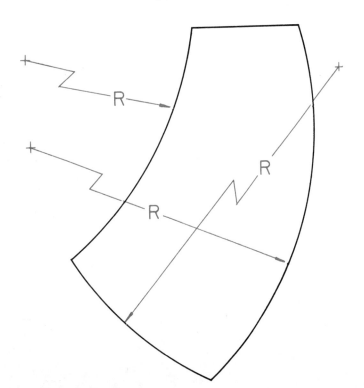

Fig. 8-1. How radii are dimensioned for outside arcs.

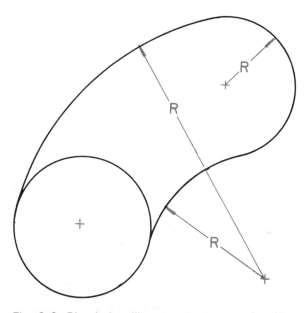

Fig. 8-3. Blended radii run tangent to each other.

drawing. A radius which has its origin outside the drawing is dimensioned as shown in Fig. 8-1.

Some radii may have a defined origin, Fig. 8-2. Others may have a "blend" radius which runs tangent with other arcs or lines. See Fig. 8-3.

FILLETS AND ROUNDS

Fillets and rounds are radii on the inside and outside corners of a workpiece. They are frequently found on castings, forgings, and stampings.

The sizes of fillets and rounds are given in terms of radius, which is specified directly on the print, Fig. 8-4, or given as a special note on the drawing, Fig. 8-5.

A FILLET is a radius put on the inside corners of a workpiece to increase its strength and appearance, Fig. 8-4.

A ROUND is a radius of the outside edges/corners of a workpiece to improve its appearance or to remove a sharp edge or corner which could be subject to breakage. See Fig. 8-4.

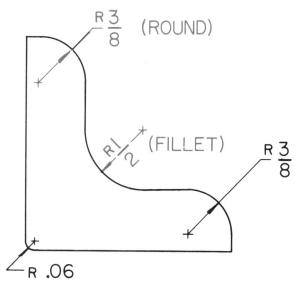

Fig. 8-4. How fillet and rounds are specified on a drawing.

NOTE:

UNLESS OTHERWISE SPECIFIED ALL ROUNDS AND FILLETS R. $\frac{1}{8}$

NOTE:

ALL RADII R .06 UNLESS NOTED.

Fig. 8-5. Special notes call out sizes of radii.

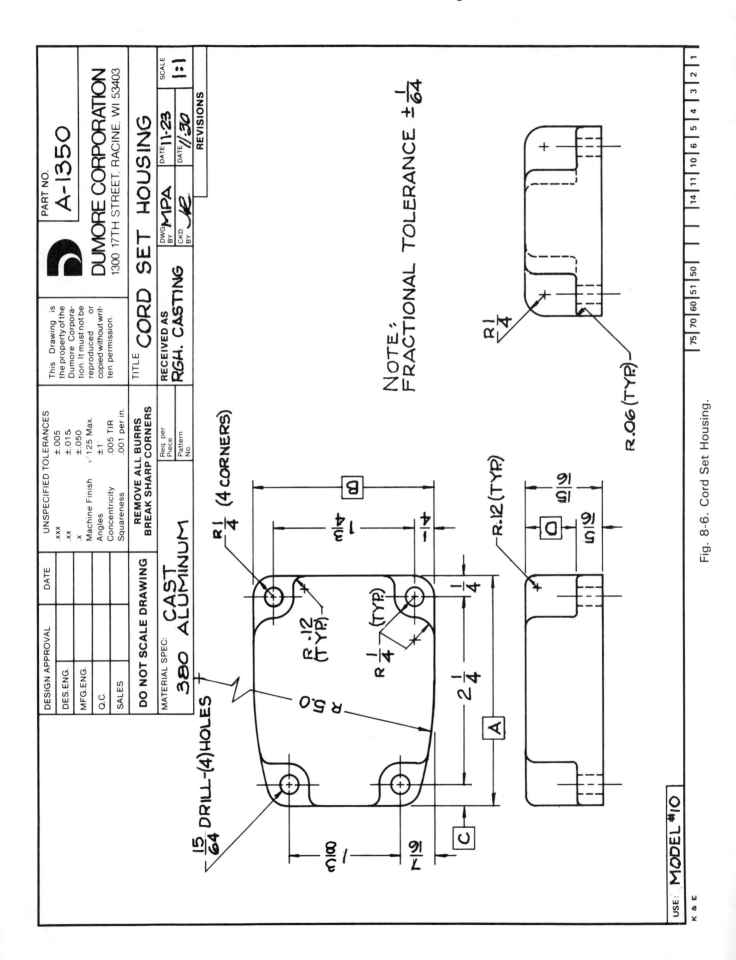

Fig. 8-6. Cord Set Housing.

DIRECTIONS—QUIZ QUESTIONS

1. The industrial prints in this section will test your print reading ability.
2. Study the views, dimensions, title block, and notes in Figs. 8-6 through 8-10.
3. Read the quiz questions, refer to the print, and write your answers in the blanks provided.

CORD SET HOUSING QUIZ

1. List the various radii found on the workpiece.

1. _____

2. Determine dimension [A] .

2. _____

3. What name is given to the R .12 corners?

3. _____

4. What kind of material is used to produce the workpiece?

4. _____

5. Determine dimension [B] .

5. _____

6. What is the overall height of the workpiece?

6. _____

7. What size is the largest radius found on the workpiece?

7. _____

8. Determine dimension [C] .

8. _____

9. How many holes does the casting require?

9. _____

10. In what condition was the material received?

10. _____

11. What tolerance is used for fractional dimensions?

11. _____

12. How many surfaces are shown in the top view?

12. _____

13. Determine distance [D] .

13. _____

14. Does the print contain any revisions?

14. _____

15. What scale is the print?

15. _____

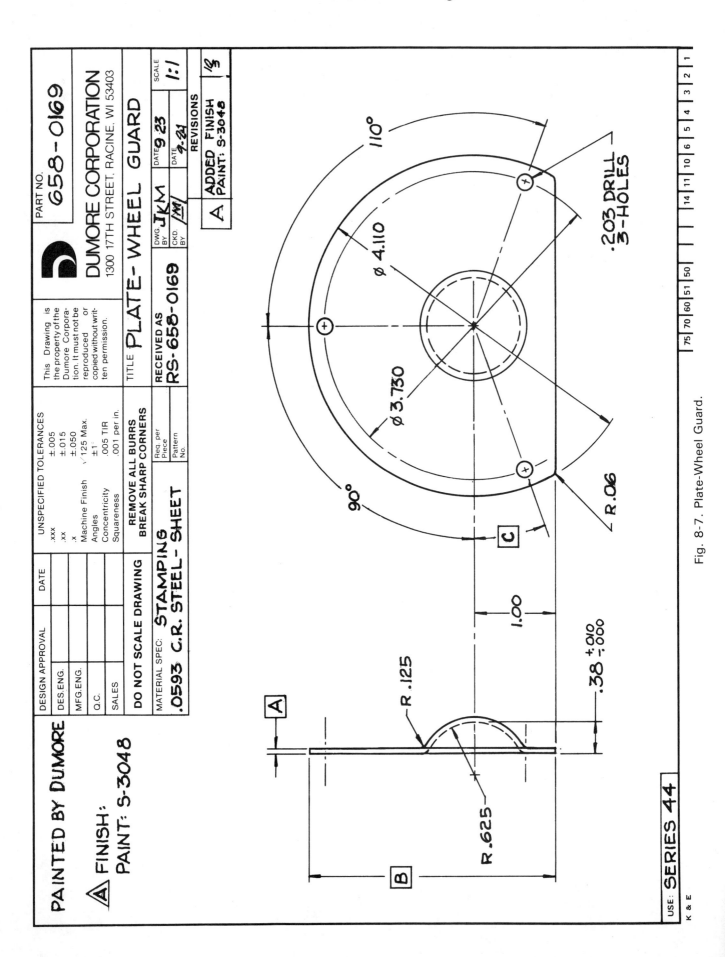

Fig. 8-7. Plate-Wheel Guard.

PLATE—WHEEL GUARD QUIZ

1. What is the part number of the print?

2. What is revision A ?

3. What is dimension A ?

4. How many holes are in the workpiece?

5. What type of tolerance is used on the .38 decimal dimension?

6. What is the outside diameter of the workpiece?

7. What kind of material is used to produce the part?

8. What is dimension B ?

9. What is radius R .125 called?

10. What is dimension C ?

11. What is the overall height of the workpiece to the outside surface of the depression?

12. What is the high limit on the 1.00 dimension?

13. How far are the holes from the center of the depression?

14. What tolerance is required on angles?

15. How is this part produced?

16. List the radii found on the workpiece.

17. Are the three holes equally spaced on a circle?

18. What note is given regarding scale?

19. What information is found on the print regarding corners?

20. What tolerance is required on three place decimal dimensions?

1. _____
2. _____
3. _____
4. _____
5. _____
6. _____
7. _____
8. _____
9. _____
10. _____
11. _____
12. _____
13. _____
14. _____
15. _____
16. _____
17. _____
18. _____
19. _____
20. _____

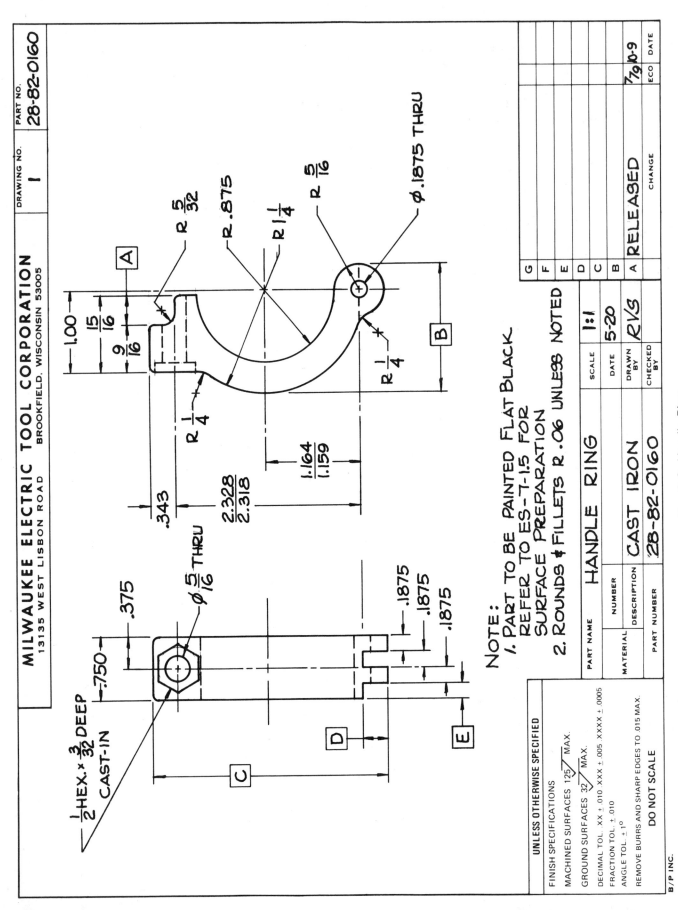

Fig. 8-8. Handle Ring.

HANDLE RING QUIZ

1. What size is the smallest diameter hole?
2. How deep is the hex counterbore?
3. Determine dimension [A] .
4. What size is the largest radius shown on the print?
5. What is the maximum allowed distance between the two holes on the part?
6. Determine distance [B] .
7. What is the part number?
8. List all the radii contained on the part.
9. What is the mean dimension on the 2.328/2.318 dimension?
10. What color is the casting painted?
11. Determine dimension [C] .
12. How thick is the casting?
13. What material is used to make the part?
14. What two views are shown on the print?
15. How far is the 3/16 diameter hole from the center of the part?
16. Determine dimension [D] .
17. What is the high limit on the .343 dimension?
18. How is the 1/2 hex counterbore made?
19. What size are unspecified fillets and rounds?
20. Determine dimension [E] .

1. _____
2. _____
3. _____
4. _____
5. _____
6. _____
7. _____
8. _____
9. _____
10. _____
11. _____
12. _____
13. _____
14. _____
15. _____
16. _____
17. _____
18. _____
19. _____
20. _____

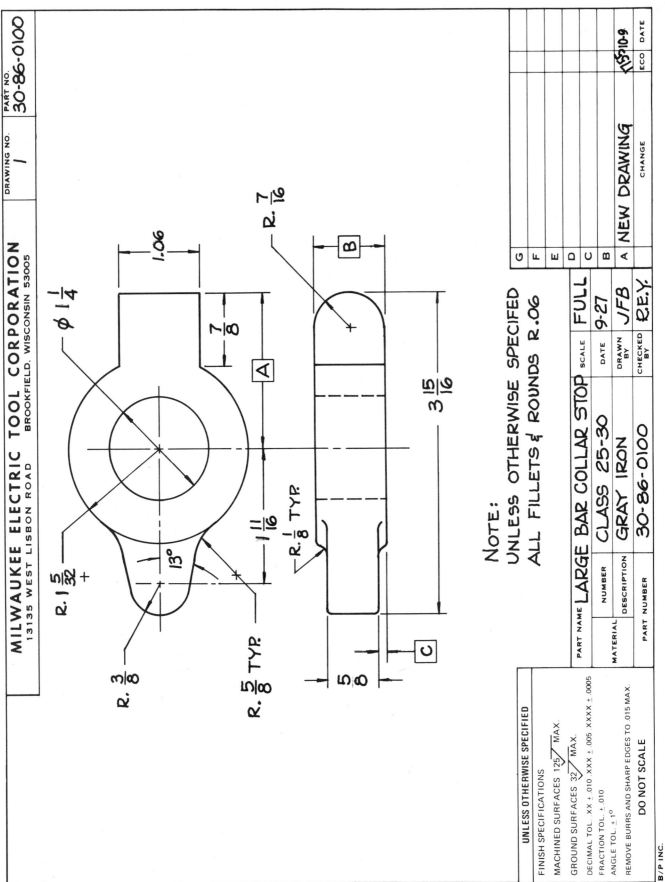

MILWAUKEE ELECTRIC TOOL CORPORATION
13135 WEST LISBON ROAD   BROOKFIELD, WISCONSIN 53005

DRAWING NO. 1
PART NO. 30-86-0100

ø 1 1/4
R. 1 5/32
R. 3/8
R. 5/8 TYP.
13°
1 11/16
R. 1/8 TYP.
1.06
7/8
A
R. 7/16
B
3 15/16
5/8
C

NOTE:
UNLESS OTHERWISE SPECIFIED
ALL FILLETS & ROUNDS R.06

PART NAME	LARGE BAR COLLAR STOP		SCALE	FULL
MATERIAL	NUMBER	CLASS 25-30	DATE	9-27
	DESCRIPTION	GRAY IRON	DRAWN BY	JFB
	PART NUMBER	30-86-0100	CHECKED BY	R.E.Y.

G		
F		
E		
D		
C		
B		
A	NEW DRAWING	9-15-10
	CHANGE	ECO DATE

UNLESS OTHERWISE SPECIFIED
FINISH SPECIFICATIONS
MACHINED SURFACES 125 MAX.
GROUND SURFACES 32 MAX.
DECIMAL TOL..XX ± .010 .XXX ± .005 .XXXX ± .0005
FRACTION TOL. ± .010
ANGLE TOL. ± 1°
REMOVE BURRS AND SHARP EDGES TO .015 MAX.
DO NOT SCALE

B/P INC.

Fig. 8-9. Large Bar Collar Stop.

LARGE BAR COLLAR STOP QUIZ

1. What size is the large diameter hole?

2. What is the high limit on the large diameter hole?

3. What kind of material is used to make the part?

4. What is the material number?

5. Determine the part finish required on machined surfaces.

6. List all the radii found on the part.

7. Determine the width of the part.

8. Calculate dimension A .

9. What is the low limit on the angle shown on the print?

10. Determine height B .

11. Give the part number of the print.

12. What is the maximum overall length allowed on the part.

13. What type of dimensions are shown on the print?

14. Determine distance C .

15. What tolerance is allowed on the 5/8 dimension?

16. Who checked the print?

17. What size are the radii not specified on the print?

18. What scale is the print?

19. What two views are shown on the print?

20. What size is the smallest fillet shown on the print?

1. _____

2. _____

3. _____

4. _____

5. _____

6. _____

7. _____

8. _____

9. _____

10. _____

11. _____

12. _____

13. _____

14. _____

15. _____

16. _____

17. _____

18. _____

19. _____

20. _____

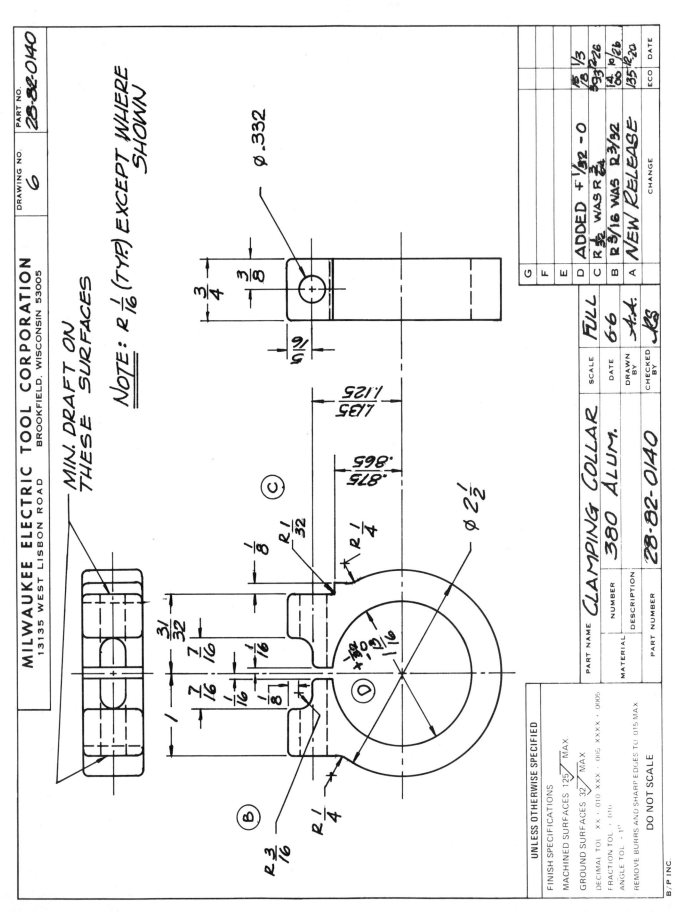

Fig. 8-10. Clamping Collar.

MILWAUKEE ELECTRIC TOOL CORPORATION
13135 WEST LISBON ROAD   BROOKFIELD, WISCONSIN 53005

DRAWING NO. 6   PART NO. 28-82-0140

MIN. DRAFT ON THESE SURFACES

NOTE: R 1/16 (TYP) EXCEPT WHERE SHOWN

ø .332

3/4   3/8

5/16

1.1125/1.135

.875/.865

C

R 1/32

1/8

R 1/4

ø 2 1/2

3/32
7/16
1/16
7/16
1/16
1/8
1

1 1/16 +.030/-.010

D

B

R 3/16

R 1/4

G				
F				
E				
D	ADDED F 1/32 - 0		15/18	1/3
C	R 5/32 WAS R 3/64		593	12-26
B	R 3/16 WAS R 3/32		14.00	10/26
A	NEW RELEASE		135	12-20
	CHANGE		ECO	DATE

PART NAME CLAMPING COLLAR   SCALE FULL
NUMBER   DATE 6-6
MATERIAL 380 ALUM.   DRAWN BY A.A.
DESCRIPTION   CHECKED BY JRS
PART NUMBER 28-82-0140

UNLESS OTHERWISE SPECIFIED

FINISH SPECIFICATIONS
MACHINED SURFACES 125 MAX.
GROUND SURFACES 32 MAX.
DECIMAL TOL   XX ·.010 XXX ·.005 XXXX ·.0005
FRACTION TOL · 1/64
ANGLE TOL · 1°
REMOVE BURRS AND SHARP EDGES TO .015 MAX

DO NOT SCALE

B/P INC.

CLAMPING COLLAR QUIZ

1. What material is used to make the part?

2. What size radii does the part have?

3. Unless otherwise specified, what are the fractional tolerances?

4. What size hole is drilled through both sides of the part?

5. How thick is the part?

6. What is the center distance between the two holes?

7. What size is the largest diameter shown on the part?

8. What scale is the print?

9. How long is the small hole?

10. What ECO number was used to designate revision B?

11. What is the high limit on size of large diameter hole?

12. What is the wall thickness between the large diameter hole and the outside diameter of the part?

13. Which two dimensions give the location of the small diameter hole?

14. What is the width of the slot which splits the clamping collar?

15. What is the part number of the print?

16. What distance is the face of the 1/8 deep flat to the vertical centerline of the part?

17. What is the tolerance on the .875/.865 dimension?

18. What size are the radii not specified on the print?

19. What specific change was made in revision C?

20. What three views are shown on the print?

1. _____

2. _____

3. _____

4. _____

5. _____

6. _____

7. _____

8. _____

9. _____

10. _____

11. _____

12. _____

13. _____

14. _____

15. _____

16. _____

17. _____

18. _____

19. _____

20. _____

Unit 9

SECTIONAL VIEWS

After studying this unit, you will be able to:
- Explain the terms: sectional view, cutting plane line, and section lines.
- Identify various section lines.
- Complete various sectional views.
- Identify sectional views such as: Full, Half, Revolved, Removed, and Broken-out.

A SECTIONAL VIEW of an object in a drawing is created by the imaginary cutting away of its front portion to reveal its interior. The exposed (cut) surface, then, is emphasized by the use of SECTION LINES (cross-hatching). See Fig. 9-1.

WORKING WITH SECTIONAL VIEWS

The use of sectional views is a graphic method of exposing the interior details of a workpiece. It is an effective way of showing inside features that would be complicated or confusing if described entirely by hidden lines. A sectional view may serve as one of the principal views—front, top, or side—on a print, or it can be used as an additional view.

A sectional view is developed by first passing an imaginary CUTTING PLANE through the workpiece. Then, the part of the workpiece nearest the "reader" is removed, thereby revealing a direct and clear view of the interior shape. See Fig. 9-1.

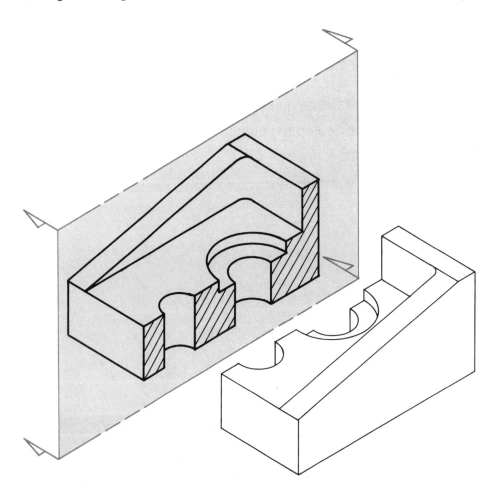

Fig. 9-1. Sectioned view shows how imaginary cutting plane cuts away front portion of part to reveal interior details.

CUTTING PLANE LINE

The location of the imaginary cutting plane is indicated by a heavy line called a CUTTING PLANE LINE, Fig. 9-2. The ends of the cutting plane line are bent at 90° with arrowheads on the ends that point in the direction of viewing sight. The side of the workpiece toward the arrows is the side that will be sectioned. The position of the cutting plane line on the part will determine the type of sectioning (full, half, revolved, removed, or broken-out).

When two or more sectional views are shown on a print, letters are placed at each end of the cutting plane line. These letters match the letters shown directly below the sectional view identified with that cutting plane line. See Fig. 9-2.

Several types of cutting plane lines are used on industrial prints. Examples are shown in Fig. 9-3.

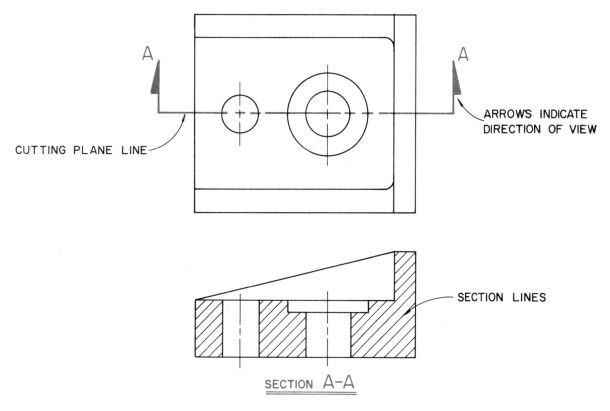

SECTION A-A

Fig. 9-2. Location of cutting plane in top view is indicated by a heavy cutting plane line marked A-A. Sectioned view below is identified as Section A-A.

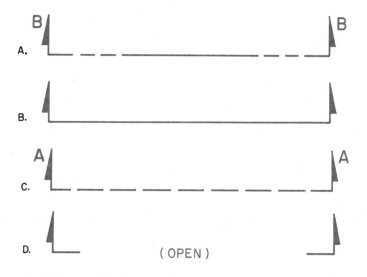

Fig. 9-3. Examples of various cutting plane lines are presented at A, B, C, and D.

SECTION LINES

SECTION LINES (cross-hatch lines) are used to identify and emphasize the surfaces that have been cut and exposed by a cutting plane line.

Section lines are thin, parallel, slanted lines, usually drawn at an angle of 45°, Fig. 9-4. However, if the sectional lines become parallel with part of the section outline, then they must be drawn at some other angle (usually 30° or 60°). Both poor practice and preferred method are shown in Fig. 9-5.

When it becomes necessary to show different kinds of materials (on assembly prints, for example), various types of section lines are used. Section lines are used on all types of materials, nonmetallic as well as metallic. See Fig. 9-6.

In general use, however, the sectional lines for cast iron are applied to working (detail) drawings of a separate part.

FULL SECTIONS

A FULL SECTION is created by passing the cutting plane line through the entire object, as shown in Fig. 9-7. Lines which were hidden are now exposed and shown as solid object lines.

NOTE: Only the edges that the cutting plane line touch are shown. The hidden lines behind the cutting plane line are generally omitted. See views A and B in Fig. 9-7. They would merely add confusion to the interior detail.

First, determine the direction of the cutting plane line on the corresponding view. Then transfer the details behind the cutting plane line to the sectioned view. The transferring of points on the cutting plane line to the sectional view locates the interior detail. Note that the outer edges of the part are always defined by solid object lines; hidden lines would not be permissible.

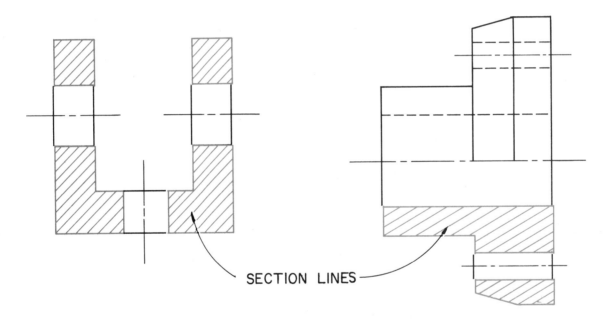

SECTION LINES

Fig. 9-4. Section lines emphasize surfaces that have been cut. Usually, they are drawn at a 45° angle.

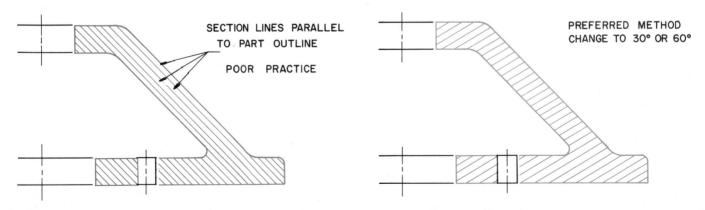

SECTION LINES PARALLEL TO PART OUTLINE

POOR PRACTICE

PREFERRED METHOD CHANGE TO 30° OR 60°

Fig. 9-5. View at top illustrates poor practice because section lines are drawn parallel with part object lines. View below shows preferred method of section lining.

STEEL

CAST IRON
OR
GENERAL USAGE

BRONZE, BRASS,
COPPER

WHITE METAL,
ZINC, LEAD,
BABBITT, ALLOYS

MAGNESIUM,
ALUMINUM, AND
ALLOYS

CORK, FELT
FABRICS, FIBER,
LEATHER

RUBBER, PLASTIC,
ELECTRICAL INSULATION

Fig. 9-6. Material symbols for sectioning are pictured and labeled.

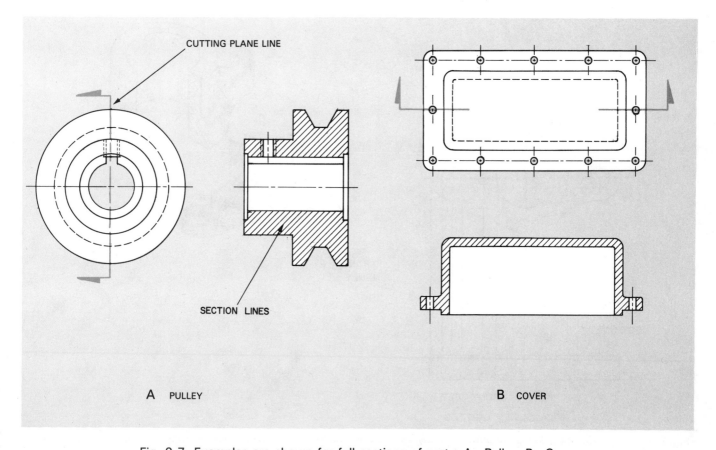

Fig. 9-7. Examples are shown for full sections of parts: A—Pulley. B—Cover.

With practice, this technique will be most helpful in visualizing the other types of sectional views.

HALF SECTIONS

A HALF SECTION of a symmetrical (both halves the same) object shows the interior and exterior features in the same view. See Fig. 9-8. One half of the view will show the sectioning of the interior of the object, while the remaining half will detail the exterior. Again, section views are drawn with solid object lines. Hidden lines may tend to add more confusion to the section lines.

The omitted hidden lines are standard practice if the part is not complicated. However, hidden lines may be of value when used with forged parts or complicated cast parts for dimensioning purposes.

The cutting plane line in Fig. 9-8 is drawn at a 90° angle toward the right side of the top view. Follow the transfer lines from the top view into the front view. The right side of the front view has been sectioned by the use of the cutting plane line and sectioning lines. The main centerline in the front view will remain unchanged as the center of the part in Fig. 9-8.

The half section can easily be seen in the front view. As in all half sections, the sectioned portion will only be one half of that view.

REVOLVED SECTION

A REVOLVED SECTION is a sectional drawing which represents a single portion of the workpiece. Revolved sections are cross sections of the shape not shown in the conventional views. See Fig. 9-9.

A centerline is drawn through the portion which will be revolved and sectioned. The centerline acts as an imaginary viewing plane (cutting plane) that revolves 90° to the original view on the centerline. The section is revolved directly on the surface of the original view and drawn into true size and shape. Keep in mind, the drafter or designer must provide the true size and shape of the contours. The lines from the view may overlap with the revolved section. These should be omitted, view A in Fig. 9-9.

Revolved sections are commonly applied to arms, ribs, spokes, bars, and irregular contoured parts.

REMOVED SECTION

A REMOVED SECTION is similar to the revolved section, but "removed" from its position within a view to a new location elsewhere on the print. The relocating of this removed section allows the drafter to enlarge that particular section for clarity. Also, by enlarging the removed section, dimensioning may be added for further clarification.

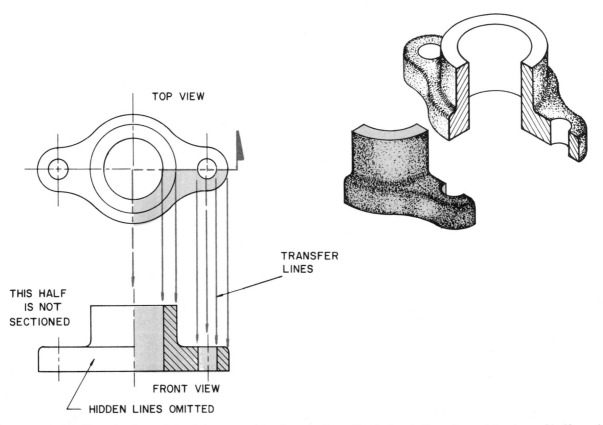

TOP VIEW

TRANSFER LINES

THIS HALF IS NOT SECTIONED

FRONT VIEW

HIDDEN LINES OMITTED

Fig. 9-8. Top view of housing is projected downward to show half section in front view. Isometric view of half section is pictured at right.

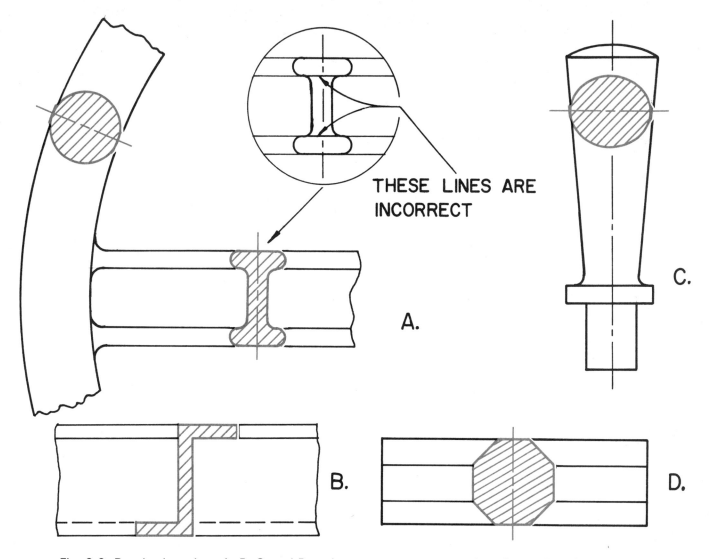

THESE LINES ARE INCORRECT

A.

B.

C.

D.

Fig. 9-9. Revolved sections A, B, C, and D each represents a cross section of a portion of workpiece.

Past practices have indicated that when more than one removed section is shown, each section should be marked separately. The removed sections are marked in alphabetical order, starting with Section A-A (then Section B-B, Section C-C, etc.) with each of the corresponding letters on the cutting plane line ends. See Fig. 9-10. The use of a cutting plane line marked with letters at both ends is the main difference between a removed section and the revolved section.

BROKEN-OUT SECTION (PARTIAL)

The BROKEN-OUT SECTION is a special type of a sectioning application, Fig. 9-11. Often, a small, single-detail portion of the part's interior needs more clarification. The broken-out section enables the print reader to concentrate directly on that detailed portion of the part. By concentrating only on that portion, the detail would be more visible to the print reader. Otherwise, the detail may be lost if a full

section or half section were applied.

A broken-out section can be located on a print by the use of a wavy line called a BREAK LINE. See views A and B in Fig. 9-11.

In this case, the break line acts as an imaginary cutting plane line. The large and heavy cutting plane line with arrows would not be placed on the drawing. The area within the break line to the part outline is section lined according to material type. Note steel section lining in view A, Fig. 9-11, and cast iron section lining in view B.

DIRECTIONS FOR SECTION PROBLEMS

The following problems will require you to sketch finished examples of the various section types. Problems include Full, Half, Revolved, Removed, and Broken-out sections, Figs. 9-12 to 9-18.

Each problem will have specific directions to follow. Use a straightedge or rule to complete the sketches.

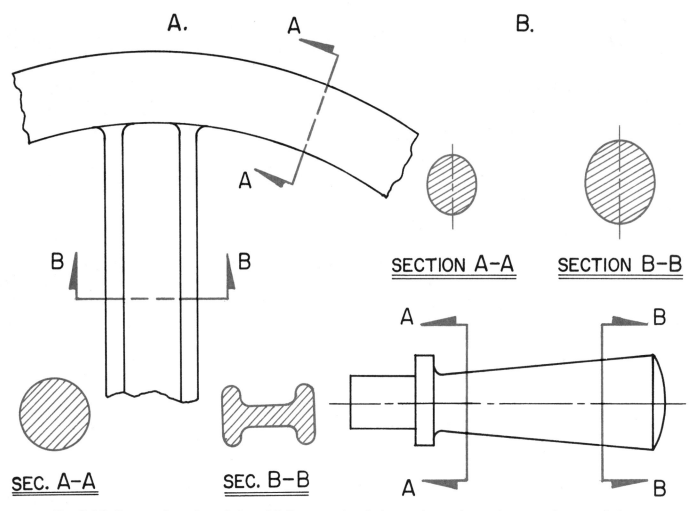

Fig. 9-10. Removed sections A-A and B-B are sectional views relocated on print away from workpiece.

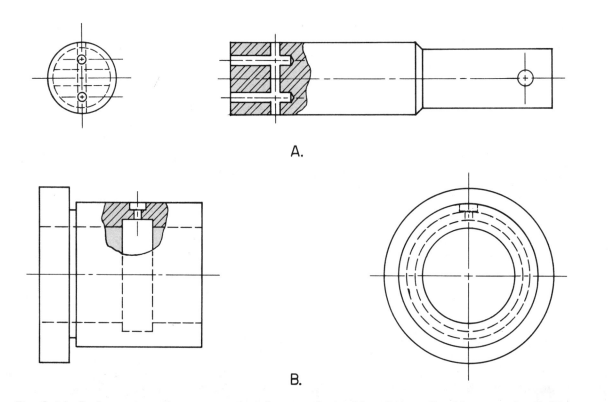

Fig. 9-11. Broken-out sections are sectioned areas of parts' interiors outlined by wavy break lines.

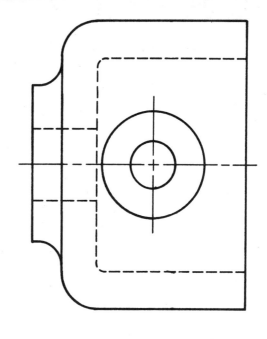

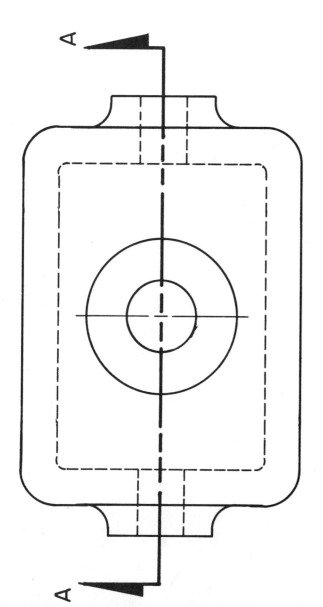

A

A

SECTION A–A

Fig. 9-12. Sectioning Problem 1: Draw front view as a full section.

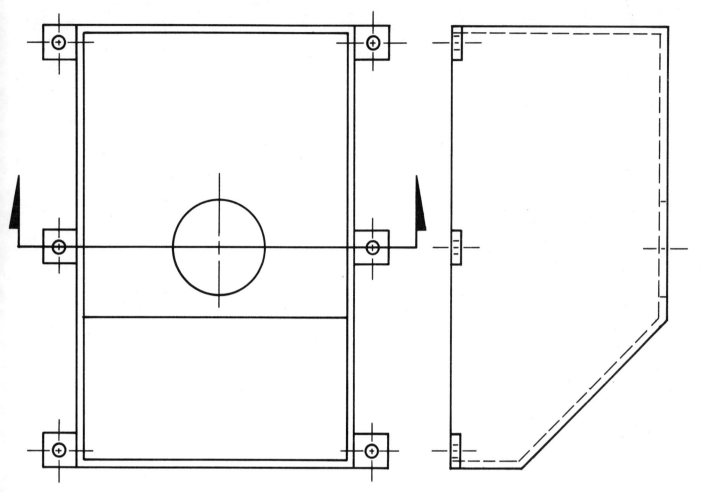

Fig. 9-13. Sectioning Problem 2: Complete left side view as a full section.

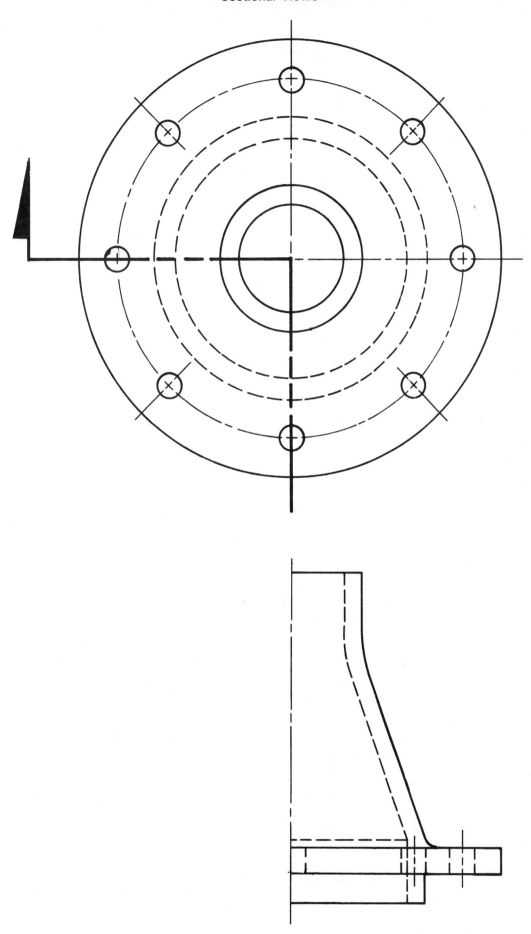

Fig. 9-14. Sectioning Problem 3: Complete this view as a half section.

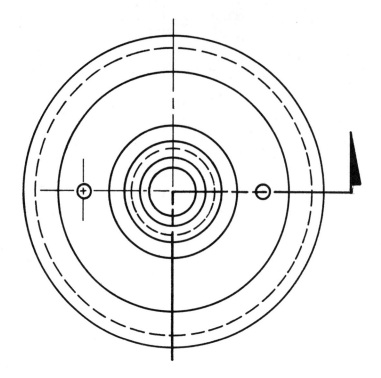

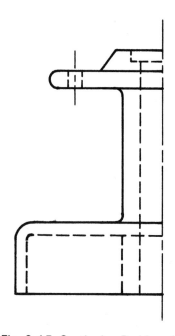

Fig. 9-15. Sectioning Problem 4: Make a half section of front view.

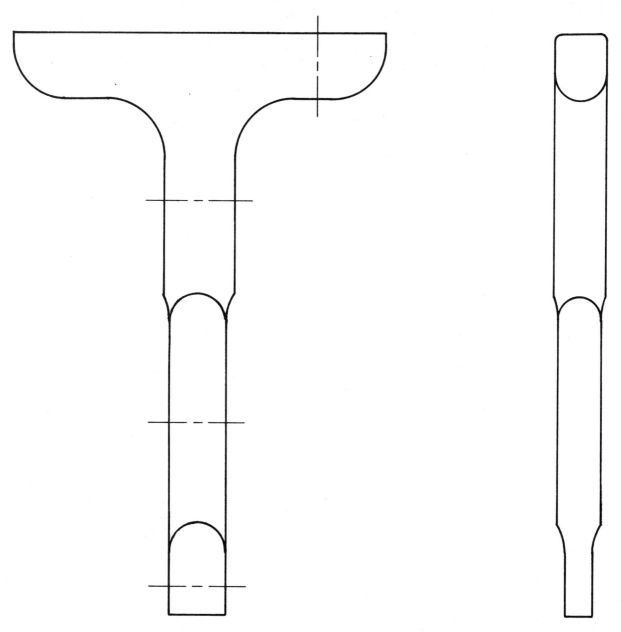

Fig. 9-16. Sectioning Problem 5: Draw revolved section on centerlines provided.

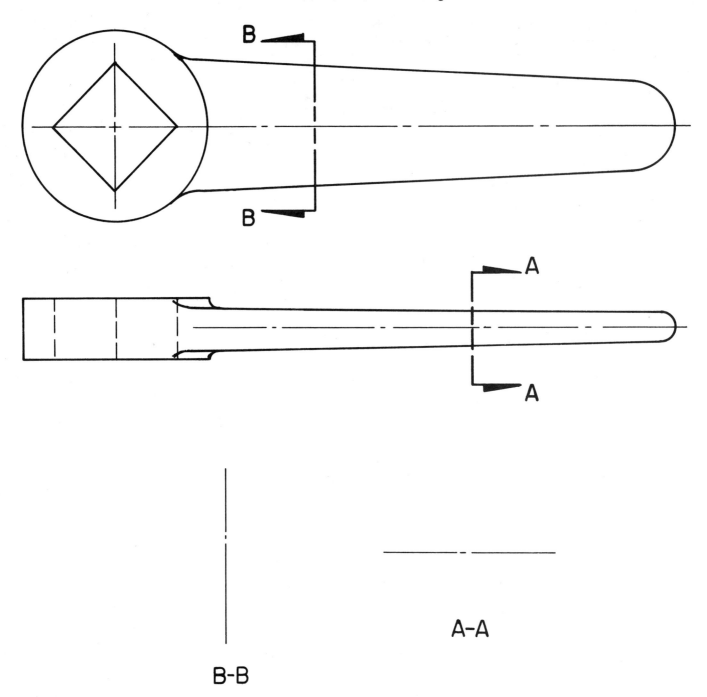

Fig. 9-17. Sectioning Problem 6: Draw removed sections on centerlines provided. Use rounded edges where necessary.

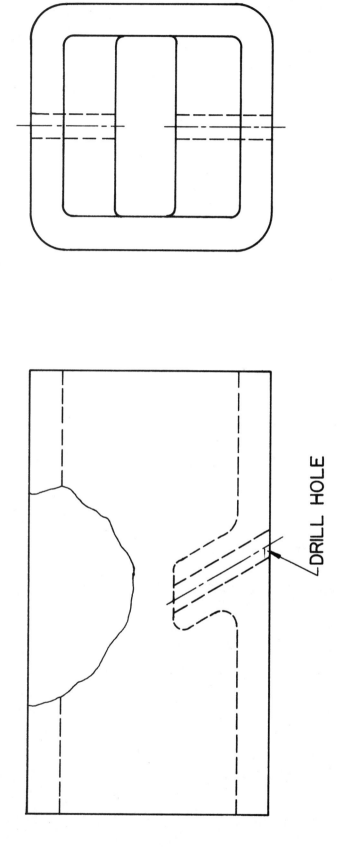

DRILL HOLE

Fig. 9-18. Sectioning Problem 7: Draw detail for broken-out section.

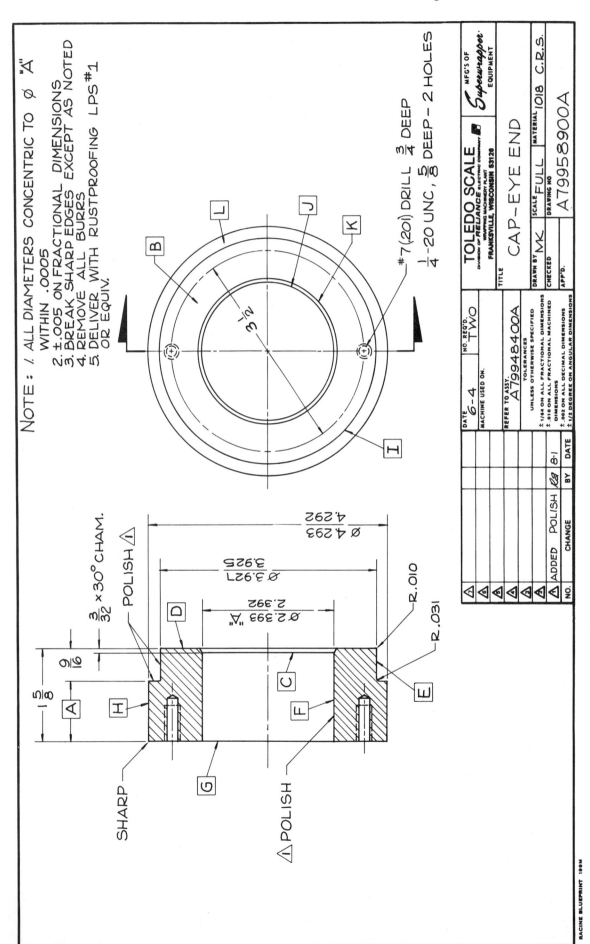

NOTE: 1. ALL DIAMETERS CONCENTRIC TO ∅ "A" WITHIN .0005
2. ±.005 ON FRACTIONAL DIMENSIONS
3. BREAK SHARP EDGES EXCEPT AS NOTED
4. REMOVE ALL BURRS
5. DELIVER WITH RUSTPROOFING LPS #1 OR EQUIV.

#7 (201) DRILL $\frac{3}{4}$ DEEP
$\frac{1}{4}$-20 UNC, $\frac{5}{8}$ DEEP – 2 HOLES

$\frac{3}{32} \times 30°$ CHAM.

POLISH

$1\frac{5}{8}$
$\frac{9}{16}$

∅ 4.293 / 4.292
∅ 3.925 / 3.921
∅ 2.393 / 2.392 "A"

R.010
R.031

SHARP

POLISH

DATE 6-4	NO. REQ'D TWO
MACHINE USED ON.	

REFER TO ASSY. A799484OOA

TOLERANCES
UNLESS OTHERWISE SPECIFIED
± 1/64 ON ALL FRACTIONAL DIMENSIONS
± .010 ON ALL FRACTIONAL MACHINED DIMENSIONS
± .002 ON ALL DECIMAL DIMENSIONS
± 1/2 DEGREE ON ANGULAR DIMENSIONS

⚠	
⚠	
⚠	
⚠	
⚠	
⚠	
⚠ ADDED POLISH RB 8-1	
NO. CHANGE BY DATE	

TOLEDO SCALE
DIVISION OF *RELIANCE* ELECTRIC COMPANY
WRAPPING MACHINERY PLANT
FRANKSVILLE, WISCONSIN 53126

MFG'S OF *Superwrapper.* EQUIPMENT

TITLE **CAP-EYE END**

DRAWN BY MK	
CHECKED	SCALE FULL	MATERIAL 1018 C.R.S.
APP'D.	DRAWING NO A79958900A

A79958900A

Fig. 9-19. Cap-Eye End.

RACINE BLUEPRINT 199M

DIRECTIONS—QUIZ QUESTIONS

1. The industrial prints in this section will test your print reading ability.
2. Study the views, dimensions, title block, and notes in Figs. 9-19 through 9-24.
3. Read the quiz questions, refer to the print, and write your answers in the blanks provided.

CAP-EYE END QUIZ

1. What kind of sectional view is shown?   1. _____

2. How many surfaces are polished?   2. _____

3. How long is the 3.927 diameter?   3. _____

4. What size tap drill is used?   4. _____

5. How deep should the holes be tapped?   5. _____

6. What assembly print is this part used on?   6. _____

7. What kind of material is used for the part?   7. _____

8. What does note 4 state?   8. _____

9. What specification is given for the rear outer edge of   9. _____
   the part?

10. What size is the bore on the part?   10. _____

11. What is the low limit on the large outside diameter   11. _____
    of the part?

12. Which diameters must be concentric to diameter "A"?   12. _____

13. What size chamfer is called for on the part? 13. _____

14. What is dimension A ?   14. _____

15. What surface or line in the side view is the same   15. _____
    as I in the front view?

16. Line D in the side view is what surface in the   16. _____
    front view?

17. How many parts are required?   17. _____

18. Line F in the side view is what line in the front   18. _____
    view?

19. Is surface G shown in the front view?   19. _____

20. What tolerance is used on fractional dimensions?   20. _____

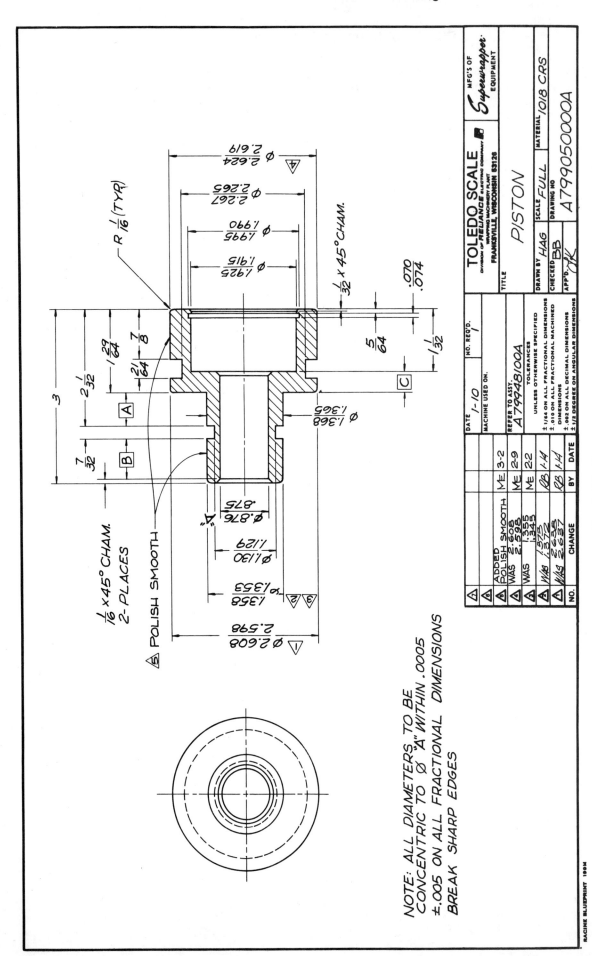

NOTE: ALL DIAMETERS TO BE
CONCENTRIC TO Ø "A" WITHIN .0005
±.005 ON ALL FRACTIONAL DIMENSIONS
BREAK SHARP EDGES

Fig. 9-20. Piston.

PISTON QUIZ

1. How wide is the internal groove in the workpiece?

2. List all the outside diameters.

3. What diameter is the internal groove?

4. How long is the 1.358/1.353 diameter?

5. How far is the 7/32 wide groove from the left end of the workpiece?

6. How long is the .875 hole (include chamfers)?

7. Determine dimension A .

8. What is the mean dimension on the counterbore diameter?

9. What are the specifications for chamfers?

10. How many pieces are required?

11. How far is the internal groove located from the right end of the workpiece?

12. How many chamfers are required on the workpiece?

13. Determine dimension B .

14. What tolerance is required on fractional dimensions?

15. How deep is the counterbore?

16. What was revision 2 ?

17. Determine dimension C .

18. What is the radius for rounds?

19. What maximum wall thickness could there be between the 7/32 wide groove and the .875 diameter?

20. Determine the distance from the shoulder formed by the 2.624 and 1.368 diameters and the left end of the workpiece.

1. _____

2. _____

3. _____

4. _____

5. _____

6. _____

7. _____

8. _____

9. _____

10. _____

11. _____

12. _____

13. _____

14. _____

15. _____

16. _____

17. _____

18. _____

19. _____

20. _____

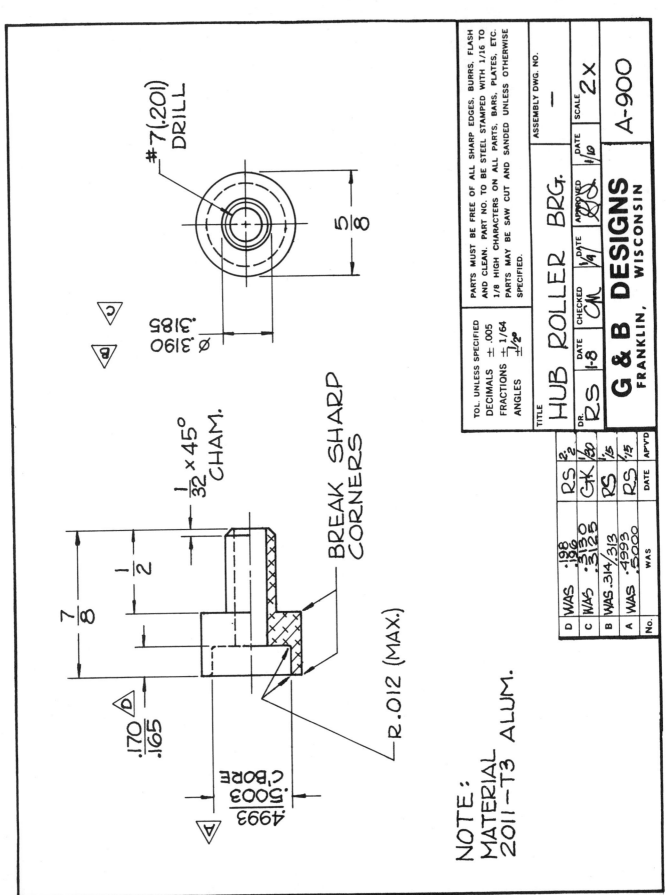

#7 (.201)
DRILL

$\frac{5}{8}$

Ø .3190
.3185

$\frac{1}{32} \times 45°$ CHAM.

$\frac{1}{2}$

$\frac{7}{8}$

BREAK SHARP
CORNERS

R .012 (MAX.)

.170
.165

.4993
.5003 C'BORE

NOTE:
MATERIAL
2011-T3 ALUM.

TOL. UNLESS SPECIFIED	PARTS MUST BE FREE OF ALL SHARP EDGES, BURRS, FLASH AND CLEAN. PART NO. TO BE STEEL STAMPED WITH 1/16 TO 1/8 HIGH CHARACTERS ON ALL PARTS, BARS, PLATES, ETC. PARTS MAY BE SAW CUT AND SANDED UNLESS OTHERWISE SPECIFIED.	ASSEMBLY DWG. NO.
DECIMALS ± .005 FRACTIONS ± 1/64 ANGLES ± 1/2°		1

TITLE					
HUB ROLLER BRG.					
DR. RS	DATE 1-8	CHECKED CM	DATE 1/4	APPROVED	DATE 1/10

G & B DESIGNS
FRANKLIN, WISCONSIN

SCALE 2x

A-900

No.			DATE	APVD
D	WAS .198 .196	RS 2/2		
C	WAS .3130 .3125	GK 1/30		
B	WAS .314/.313	RS 1/15		
A	WAS .4993 .5000	RS 1/15		
	WAS		DATE	APVD

Fig. 9-21. Hub Roller Bearing.

HUB ROLLER BEARING QUIZ

1. How many corners are broken?

2. How deep is the counterbore?

3. What was revision B?

4. What scale is the print?

5. What does the scale of the print actually mean?

6. What tolerance is used on unspecified decimal dimensions?

7. What size is the drilled hole?

8. What kind of sectional view is shown?

9. How long is the 5/8 diameter?

10. How long is the 45° chamfer?

11. What size radii are specified on the print?

12. What type of material is 2011-T3?

13. What is the print number?

14. What is the minimum overall length allowed on the part?

15. What was revision D?

16. How much tolerance is given on the small outside diameter?

17. What is the high limit on the counterbore diameter?

18. What do the hidden lines in the right-side view represent?

19. What shape is the part?

20. What chamfer is called for on the print?

1. _____

2. _____

3. _____

4. _____

5. _____

6. _____

7. _____

8. _____

9. _____

10. _____

11. _____

12. _____

13. _____

14. _____

15. _____

16. _____

17. _____

18. _____

19. _____

20. _____

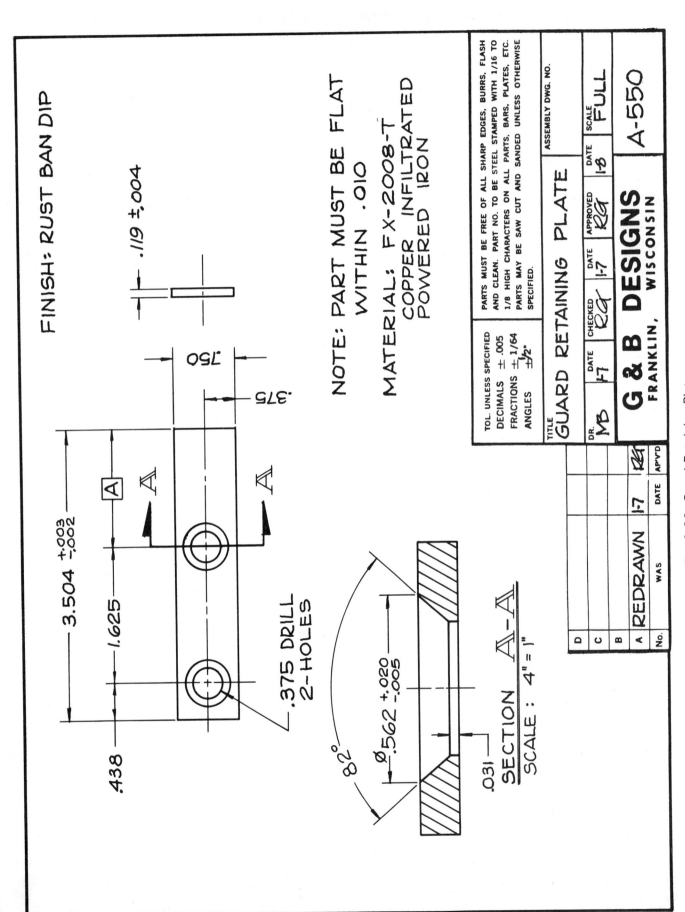

FINISH: RUST BAN DIP

.119 ±.004

NOTE: PART MUST BE FLAT
WITHIN .010

MATERIAL: FX-2008-T
COPPER INFILTRATED
POWERED IRON

.750

.375

3.504 +.003 -.002

A

1.625

.375 DRILL
2-HOLES

.438

82°

Ø.562 +.020 -.005

.031

SECTION A-A
SCALE : 4" = 1"

TOL. UNLESS SPECIFIED			PARTS MUST BE FREE OF ALL SHARP EDGES, BURRS, FLASH AND CLEAN. PART NO. TO BE STEEL STAMPED WITH 1/16 TO 1/8 HIGH CHARACTERS ON ALL PARTS, BARS, PLATES, ETC. PARTS MAY BE SAW CUT AND SANDED UNLESS OTHERWISE SPECIFIED.		ASSEMBLY DWG. NO.
DECIMALS ±.005					
FRACTIONS ±1/64					
ANGLES ±½°					
TITLE GUARD RETAINING PLATE					
DR. MB	DATE I-7	CHECKED RG	DATE I-7	APPROVED RG	DATE I-8
					SCALE FULL
G & B DESIGNS FRANKLIN, WISCONSIN					A-550

D				
C				
B				
A	REDRAWN	I-7		
No.		WAS	DATE	APV'D

Fig. 9-22. Guard Retaining Plate.

GUARD RETAINING PLATE QUIZ

1. What is the maximum length of the part?

    1. _____

2. What is the distance between the two countersunk holes?

    2. _____

3. How thick is the part?

    3. _____

4. How deep are the countersunk holes?

    4. _____

5. What is the drawing number?

    5. _____

6. What scale is the print?

    6. _____

7. What is the high limit on the countersink angles?

    7. _____

8. How wide is the part?

    8. _____

9. Calculate dimension [A]

    9. _____

10. Are any unilateral decimal tolerances shown?

    10. _____

11. What kind of sectional view is shown?

    11. _____

12. Why is the sectional view so large in comparison to the other views?

    12. _____

13. What kind of material is used for the part?

    13. _____

14. Is a surface finish required?

    14. _____

15. What is the maximum distance allowed between the two countersunk holes?

    15. _____

16. Were any revisions made to the drawing?

    16. _____

17. What size are the two drilled holes?

    17. _____

18. If the part is made to a thickness of .116 is the part over, under, or within allowable tolerance?

    18. _____

19. What size is the large diameter of the countersink?

    19. _____

20. How flat must the part be?

    20. _____

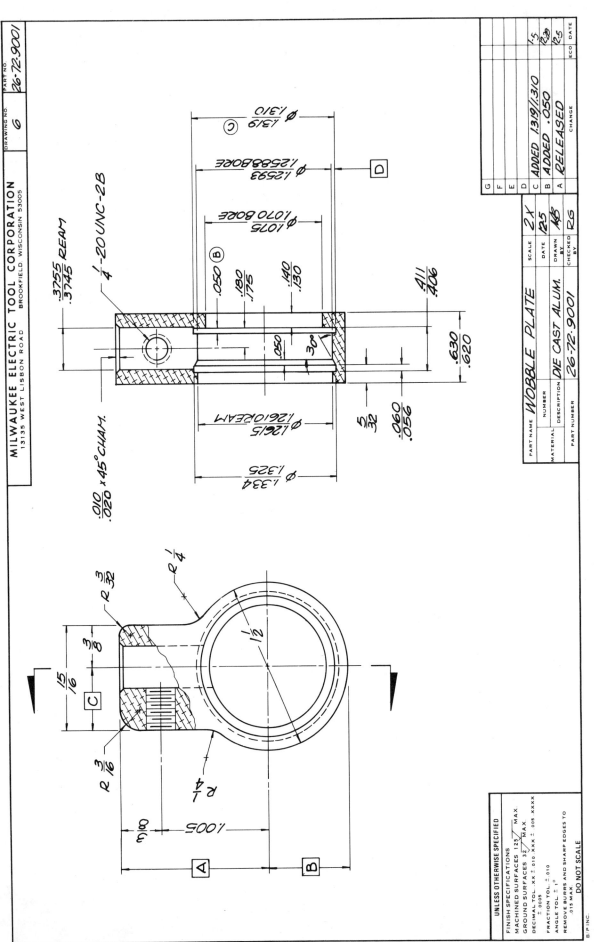

Fig. 9-23. Wobble Plate.

WOBBLE PLATE QUIZ

1. How deep is the 1.2615/1.2610 reamed hole?
1. _____

2. List the diameters of all grooves.
2. _____

3. How long is the 1.075/1.070 diameter bore?
3. _____

4. Determine distance A .
4. _____

5. What kind of material is used for the part?
5. _____

6. What kind of sectional views are used on this print?
6. _____

7. Which decimal diameter on the print allows the most tolerance?
7. _____

8. How many reamed holes are asked for on the print?
8. _____

9. What is the thread size of the tapped hole?
9. _____

10. What is the maximum allowable size (high limit) for the 1.005 dimension?
10. _____

11. List the fillets shown on the print.
11. _____

12. Determine dimension B .
12. _____

13. How long is the 30° angle (chamfer)?
13. _____

14. How long is the 1.2593/1.2588 diameter bore?
14. _____

15. What scale is the print?
15. _____

16. What class fit is the tapped hole?
16. _____

17. Determine dimension C .
17. _____

18. How much wall thickness is there between the outside diameter of the part and the diameter of the narrow groove?
18. _____

19. Determine distance D .
19. _____

20. What finish is required on the part?
20. _____

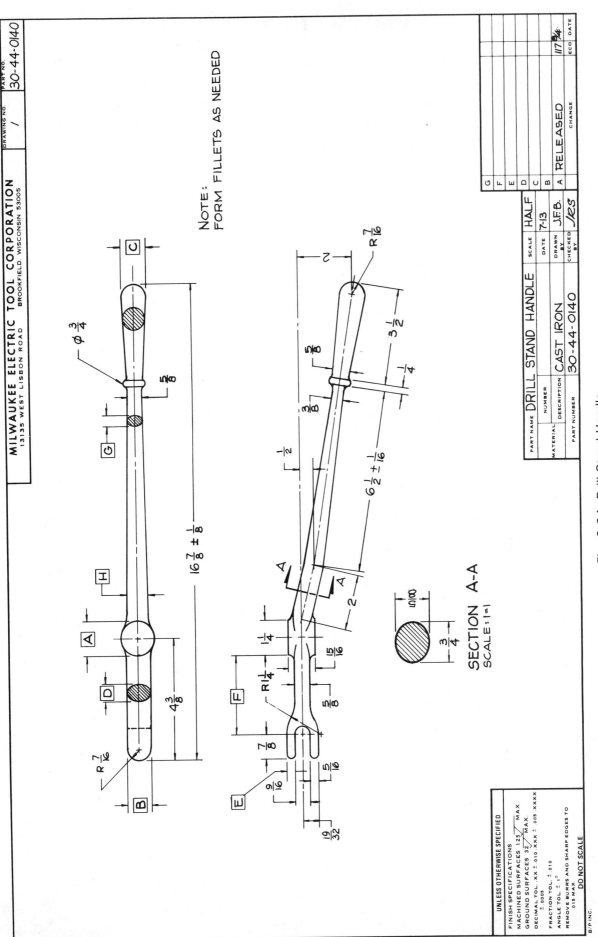

Fig. 9-24. Drill Stand Handle.

DRILL STAND HANDLE QUIZ

1. What is the maximum overall length of the workpiece?
2. How wide is the slot on the fork end of the handle?
3. What kind of sectional views are shown on this print?
4. What is dimension [A] ?
5. What scale is the print?
6. How deep is the slot?
7. What is dimension [B] ?
8. How long or wide is the 3/4 diameter shown in the top view?
9. What type of material is used to make the drill stand handle?
10. What is dimension [C] ?
11. How long is the short angular bend on the handle?
12. What is dimension [D] ?
13. What is the radius at the end of the handle?
14. Is Section A-A the same scale as the print?
15. What is dimension [E] ?
16. What is the total drop on the handle from the main centerline?
17. What is dimension [F] ?
18. What is dimension [G] ?
19. What is dimension [H] ?
20. What tolerance is used on fractional dimensions?

1. _____
2. _____
3. _____
4. _____
5. _____
6. _____
7. _____
8. _____
9. _____
10. _____
11. _____
12. _____
13. _____
14. _____
15. _____
16. _____
17. _____
18. _____
19. _____
20. _____

Unit 10

ANGLES

After studying this unit, you will be able to:
☐ Identify a chamfer and interpret its size.
☐ Define the term "taper" and interpret the meaning of TPF and TPI.
☐ Perform calculations using taper data.
☐ Define the term "bevel."
☐ Define the term "chamfer."

An ANGLE is the amount of opening between two lines that converge (meet) at a point called the "vertex." ANGULAR MEASUREMENT is based on the subdivision of a circle which has 360 degrees.

Applied to workpieces, an angle is the measurement between planes (surfaces) or locations (such as holes located on a circle) measured in degrees. Refer to Unit 4 for a further explanation of angles.

Angles found on workpieces are usually referred to as chamfers, bevels, and tapers.

CHAMFERS

A CHAMFER Is a beveled edge or angle applied to a hole or a shaft to remove sharp edges and to aid in the assembly of parts. Chamfers are often used on the ends of threaded fasteners. The chamfer aids in starting the fastener in the threaded hole, as well as serving to protect the start of the thread.

A chamfer on a hole is dimensioned by stating its diameter and angle. Fig. 10-1 shows two methods for dimensioning a chamfered hole. A chamfer on a shaft is dimensioned by stating its length and angle, as shown in Fig. 10-2.

Chamfers which are 45 degrees are dimensioned by using a leader and/or note. The term "chamfer" is optional in the note. See Fig. 10-3.

Chamfers are never measured or dimensioned

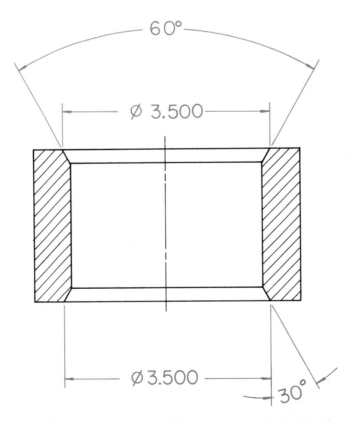

Fig. 10-1. Two methods are presented for dimensioning a chamfered hole.

128

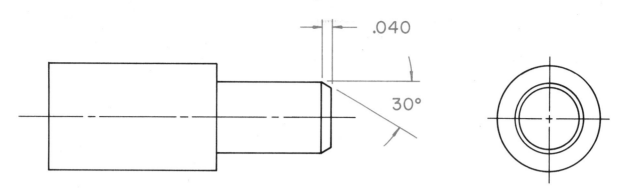

Fig. 10-2. A shaft chamfer is dimensioned by length and angle.

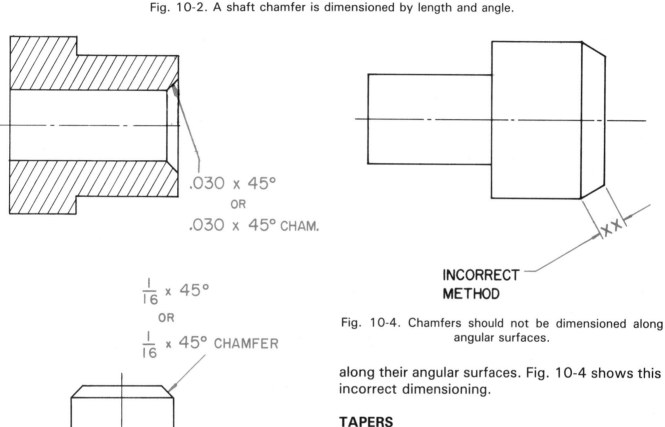

.030 x 45°
OR
.030 x 45° CHAM.

$\frac{1}{16}$ x 45°
OR
$\frac{1}{16}$ x 45° CHAMFER

INCORRECT
METHOD

Fig. 10-4. Chamfers should not be dimensioned along angular surfaces.

Fig. 10-3. A 45° chamfer is dimensioned by either method shown.

along their angular surfaces. Fig. 10-4 shows this incorrect dimensioning.

TAPERS

A conical surface on a shaft or in a hole is referred to as a TAPER. See Figs. 10-5 and 10-6. A taper uniformly changes in size along its length. Taper is often defined as the difference in diameters per unit of length (usually per in. or per ft.).

Standard machine tapers are used on cutting tools, tool holders, and machine tool spindles. Some standard machine tapers include Morse (5/8 in. taper per ft.), Brown and Sharp (1/2 in. taper per ft.), and American National Standard (3 1/2 in. taper per ft.).

Tapers can be dimensioned in several ways:
1. By stating large diameter, taper length, and including a note for standard tapers such as No. 5 Morse taper. See Fig. 10-5.
2. By indicating diameter of both ends, including taper length, Fig. 10-6.
3. By giving diameter at large end, length of taper,

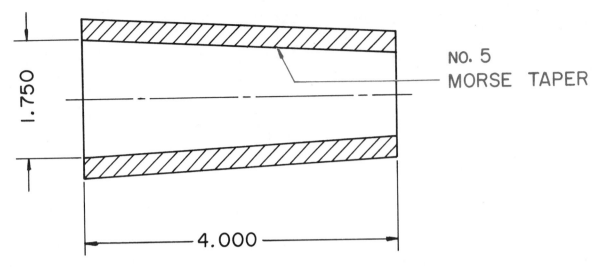

Fig. 10-5. The note, No. 5 Morse Taper, concerns a standard machine taper of 5/8 inch taper per foot (TPF).

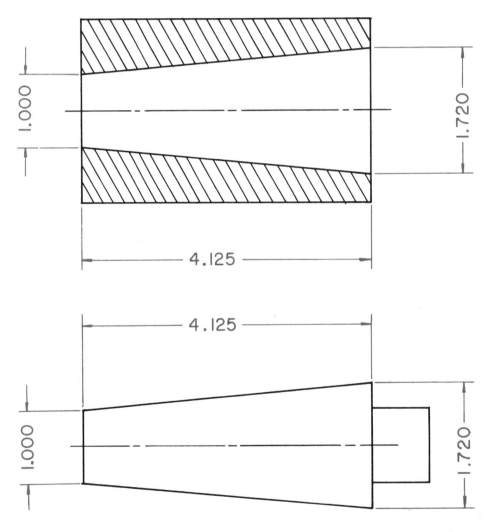

Fig. 10-6. Tapers are often dimensioned by providing large diameter, small diameter, and length of taper.

and included angle, as shown in Fig. 10-7.

4. By specifying TPI (taper per in.) or TPF (taper per ft.), taper length, and large diameter. See Fig. 10-8.

TAPER CALCULATIONS

A. Change TPI (taper per in.) to TPF (taper per ft.).
  1. Multiply TPI by 12

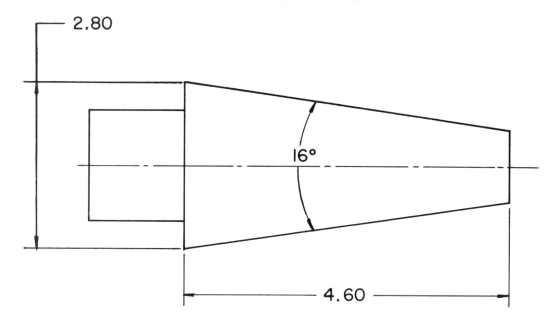

Fig. 10-7. Tapers can also be dimensioned by giving large diameter, length of taper, and included angle.

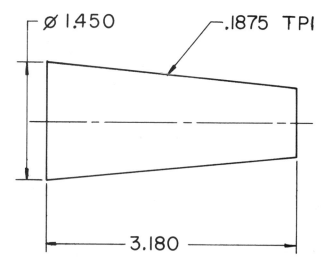

Fig. 10-8. Specifying large diameter, length of taper, and taper per inch (TPI) or taper per foot (TPF) is another acceptable way of dimensioning taper.

Example: Change .125 TPI to TPF
.125 x 12 = 1.5 in. TPF

B. Change TPF to TPI.
1. Divide TPF by 12
Example: Change 2.4 TPF to TPI
2.4 ÷ 12 = .200 in. TPI

C. Calculate small diameter, given TPI, taper length, and large diameter. See Fig. 10-9.
1. Multiply TPI by taper length.
.0625 x 1.750 = .109375 in.
2. Subtract product from large diameter
1.125 − .109375 = 1.015625 in. small diameter

D. Calculate small diameter, given taper per ft. (TPF), taper length, and large diameter. See Fig. 10-10.
1. Divide TPF by 12 to obtain TPI
.600 ÷ 12 = .050 in. TPI
2. Multiply TPI by taper length
.050 x 2.125 = .10625 in.
3. Subtract product from large diameter
1.250 − .10625 = 1.14375 in. small diameter

E. Calculate TPI, given large diameter, small diameter, and taper length. See Fig. 10-11.
1. Subtract small diameter from large diameter
1.1875 − .875 = .3125 in.
2. Divide difference by taper length
.3125 ÷ 2.125 = .147 in. TPI

BEVEL

A BEVEL is the angle one surface makes with another surface when they are not at right angles. It is the slant or inclination of a surface, Fig. 10-12.

The term "bevel" is used to denote an angle machined on a workpiece that is not cylindrical. An angle applied to a cylindrical piece or hole is referred to as a CHAMFER. Sometimes, however, the terms bevel and chamfer are used interchangeably.

ANGLES

Two methods are used to dimension angles: the coordinate method and the angular method.

The coordinate method, Fig. 10-13, uses two linear dimensions to define an angle. The coordinate

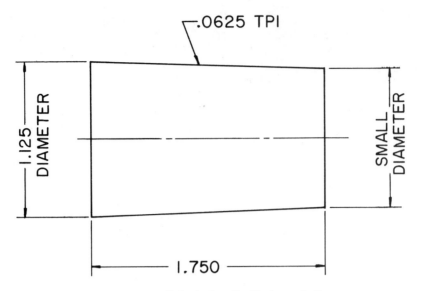

Fig. 10-9. Taper Calculation C: Find small diameter.

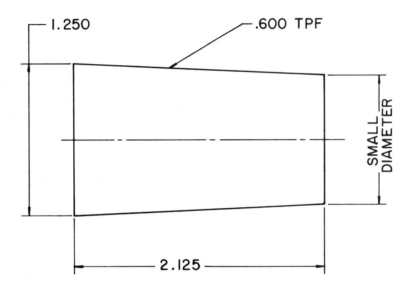

Fig. 10-10. Taper Calculation D: Find small diameter from taper per foot (TPF).

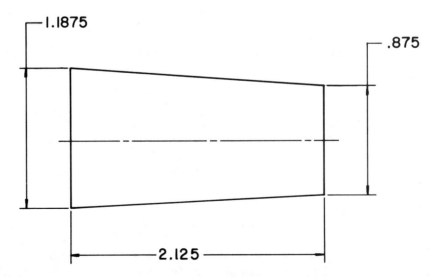

Fig. 10-11. Taper Calculation E: Find taper per inch (TPI).

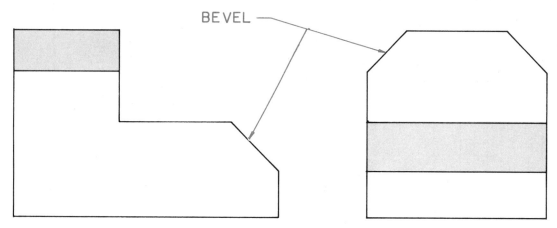

Fig. 10-12. Two view drawing shows front and side bevels.

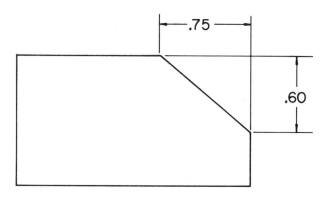

Fig. 10-13. Coordinate method of dimensioning angles.

method is used when a high degree of accuracy is desired.

The angular method, Fig. 10-14, uses one linear dimension and one angle dimension in degrees to define an angle.

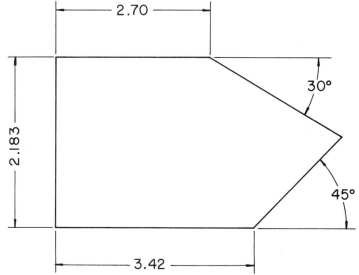

Fig. 10-14. Angular method of dimensioning angles.

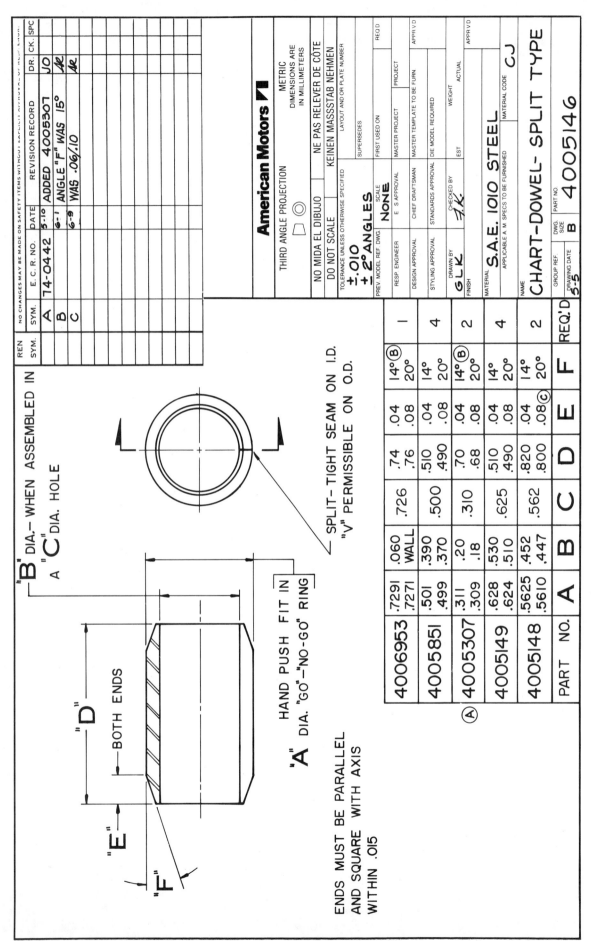

Fig. 10-15. Chart-Dowel—Split Type.

DIRECTIONS—QUIZ QUESTIONS

1. The industrial prints in this section will test your print reading ability.
2. Study the views, dimensions, title block, and notes in Figs. 10-15 through 10-19.
3. Read the quiz questions, refer to the print, and write your answers in the blanks provided.

CHART-DOWEL—SPLIT TYPE QUIZ

1. Which part has the longest length?

1. _____

2. Which part has the smallest tolerance for the outside diameter?

2. _____

3. Which size will diameter ''B'' become when assembled in a .500 diameter hole?

3. _____

4. Which part number requires 4 parts?

4. _____

5. What was revision (C) ?

5. _____

6. What material is used for the split dowels?

6. _____

7. What tolerance is used on angular dimensions?

7. _____

8. What is the high limit for ''F'' on Part No. 4005148?

8. _____

9. If Part No. 4006953 has an outside diameter of .7291, what size will diameter ''B'' become when assembled in a .726 diameter hole?

9. _____

10. How is diameter A checked?

10. _____

11. What scale is the print?

11. _____

12. What does E.C.R. mean?

12. _____

13. What is the largest tolerance given for diameter A among the various parts?

13. _____

14. Why is one half of the front view sectioned?

14. _____

15. What size is the drawing?

15. _____

16. How square must the ends of the dowel be with the axis?

16. _____

17. If the chamfer lengths of Part No. 4006953 are at the high limit, and the total length of the dowel at low limit, how long will the bearing surface of the dowel be?

17. _____

18. If the chamfer lengths of Part No. 4005851 are at the low limit, and the total length of the dowel at low limit, how long will the bearing surface of the dowel be?

18. _____

19. What is the maximum total length allowed for Part No. 4005149?

19. _____

20. Which part number was added last to the print?

20. _____

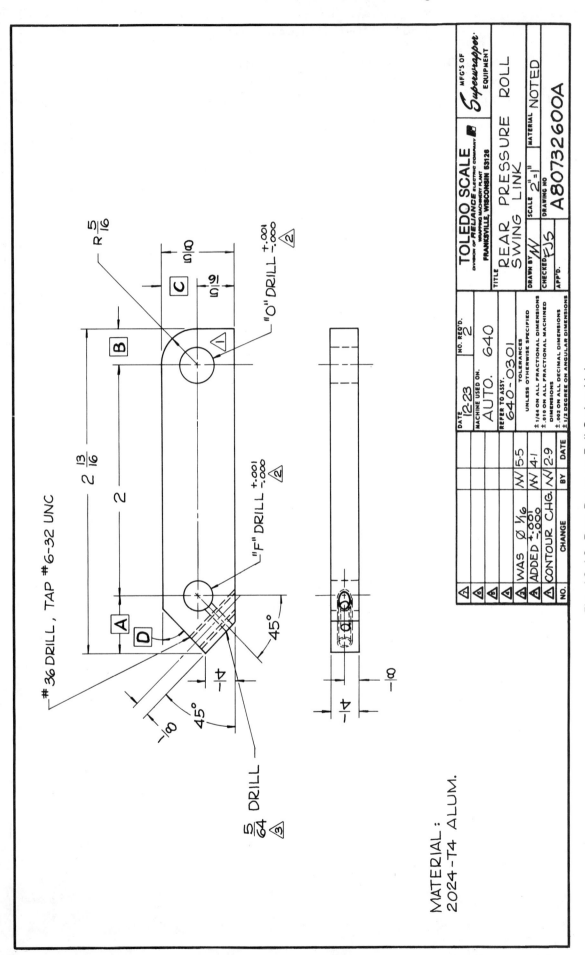

Fig. 10-16. Rear Pressure Roll Swing Link.

REAR PRESSURE ROLL SWING LINK QUIZ

1. What is the decimal size of the "O" drill?

1. _____

2. What kind of material is used to make the part?

2. _____

3. How thick is the part?

3. _____

4. How many parts are required?

4. _____

5. Determine distance ⬚B.

5. _____

6. How many holes are drilled into the part?

6. _____

7. Determine distance ⬚A.

7. _____

8. On what machine is this part used?

8. _____

9. What was revision △1 ?

9. _____

10. What tolerance is used on all decimal dimensions?

10. _____

11. What decimal size is the No. 36 tap drill?

11. _____

12. Determine distance ⬚C.

12. _____

13. Who checked the print?

13. _____

14. What tolerance is used on angular dimensions?

14. _____

15. Is the 5/64 diameter hole drilled through the center of the tapped hole?

15. _____

16. What assembly print contains this part?

16. _____

17. Determine angle ⬚D.

17. _____

18. How far from the end of the part is the tapped hole located?

18. _____

19. What is the center distance between the two largest drilled holes?

19. _____

20. At what angle is the 5/64 diameter hole drilled to the "F" hole?

20. _____

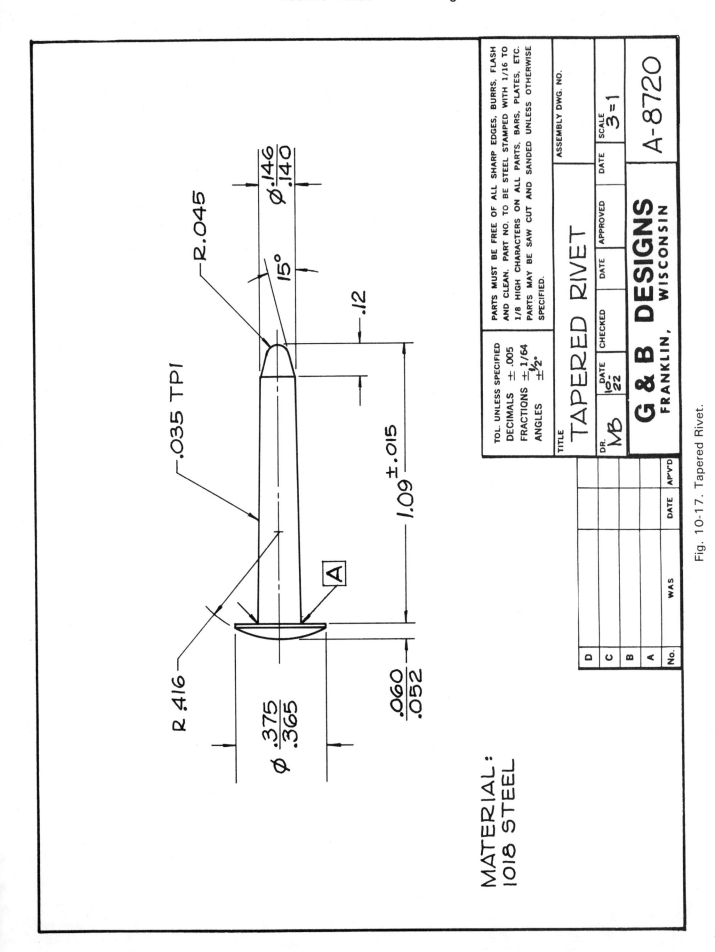

MATERIAL:
1018 STEEL

Ø .375/.365

R .416

.035 TPI

R .045

Ø .146/.140

15°

.12

1.09 ±.015

.060/.052

A

TOL. UNLESS SPECIFIED	PARTS MUST BE FREE OF ALL SHARP EDGES, BURRS, FLASH
DECIMALS ± .005	AND CLEAN. PART NO. TO BE STEEL STAMPED WITH 1/16 TO
FRACTIONS ± 1/64	1/8 HIGH CHARACTERS ON ALL PARTS, BARS, PLATES, ETC.
ANGLES ± ½°	PARTS MAY BE SAW CUT AND SANDED UNLESS OTHERWISE SPECIFIED.

				ASSEMBLY DWG. NO.	
TITLE					
TAPERED RIVET					

DR.	DATE	CHECKED	DATE	APPROVED	DATE	SCALE
MB	10-22					3=1

G & B DESIGNS
FRANKLIN, WISCONSIN

A-8720

No.	WAS		DATE	AP'VD
A				
B				
C				
D				

Fig. 10-17. Tapered Rivet.

TAPERED RIVET QUIZ

1. Determine the overall length of the rivet.

2. What diameter is the head of the rivet?

3. List the lower limit dimensions found on the print.

4. What tolerance is used for two place decimal dimensions?

5. What is the approximate large diameter of the taper behind the head of the rivet at ⬛A ?

6. What radius is the head of the rivet?

7. How much tolerance is allowed for the small diameter on the tapered rivet?

8. What is the scale of the print?

9. What kind of decimal tolerance is found on the 1.09 dimension?

10. How long is the 15° taper?

11. What finish is required on the tapered rivet?

12. What taper per inch is specified?

13. What kind of material is used to make the part?

14. What is the high limit on the 15° taper?

15. How long is the rivet head?

1. _____

2. _____

3. _____

4. _____

5. _____

6. _____

7. _____

8. _____

9. _____

10. _____

11. _____

12. _____

13. _____

14. _____

15. _____

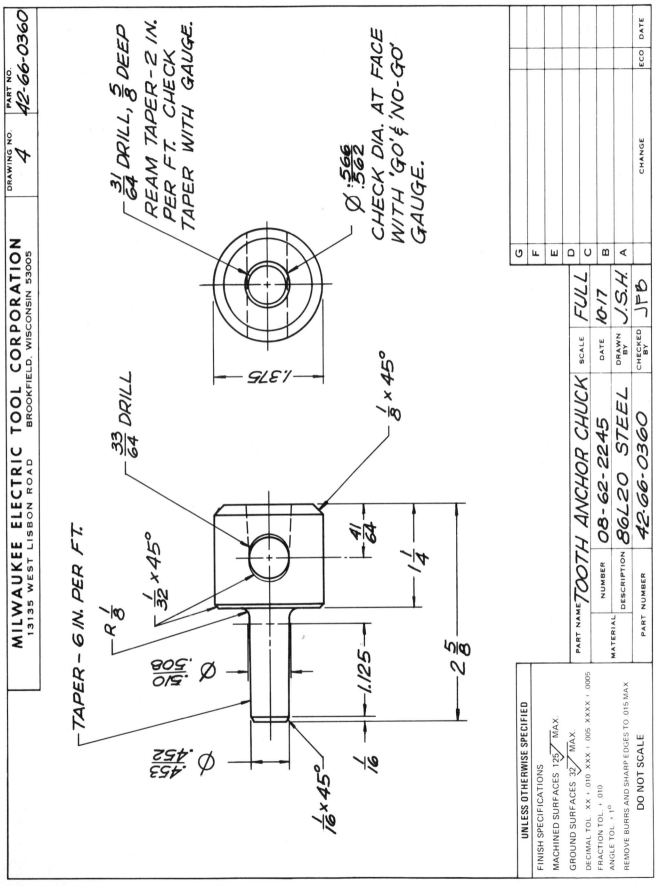

Fig. 10-18. Tooth Anchor Chuck.

TOOTH ANCHOR CHUCK QUIZ

1. List the chamfers found on the print.　　　　　　1. _____

2. What is the large diameter of the internal taper?　　2. _____

3. What size is the large end of the tapered shank?　　3. _____

4. What is the taper per inch of the chuck shank?　　4. _____

5. What is the low limit for the total length of the workpiece?　　5. _____

6. How far is the 33/64 hole from the right end of the workpiece?　　6. _____

7. What size is the small end of the tapered shank?　　7. _____

8. How deep must you drill the 31/64 hole?　　8. _____

9. How is the inside diameter at the face of the chuck checked?　　9. _____

10. What kind of material is used for the part?　　10. _____

11. What is the total length of the shank?　　11. _____

12. What tolerance is applied to the 1.125 dimension?　　12. _____

13. What is the high limit for the 33/64 hole?　　13. _____

14. What is the taper per inch of the internal hole?　　14. _____

15. What scale is the print?　　15. _____

16. What size chamfer is put on the 33/64 diameter hole?　　16. _____

17. What size radius is found on the print?　　17. _____

18. What is the material number of the part?　　18. _____

19. Who checked the print?　　19. _____

20. What is the low limit dimension of the chuck body?　　20. _____

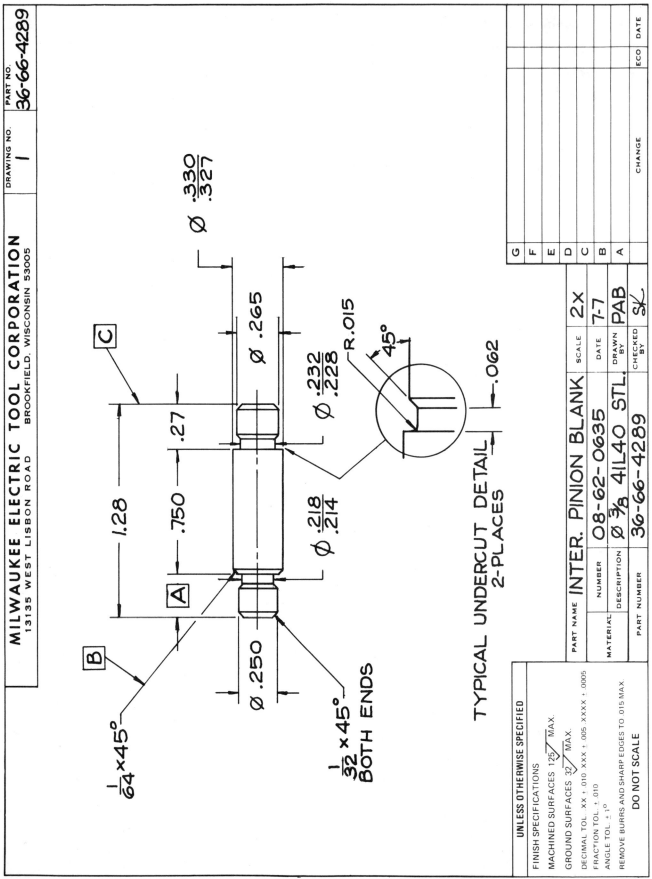

Fig. 10-19. Inter. Pinion Blank.

MILWAUKEE ELECTRIC TOOL CORPORATION
13135 WEST LISBON ROAD   BROOKFIELD, WISCONSIN 53005

DRAWING NO. 1

PART NO. 36-66-4289

Ø .330/.327

Ø .265

Ø .232/.228

R.015

45°

.062

TYPICAL UNDERCUT DETAIL
2-PLACES

.27

1.28

.750

Ø .218/.214

Ø .250

1/32 × 45°
BOTH ENDS

1/64 × 45°

PART NAME	INTER. PINION BLANK	SCALE	2X
	NUMBER	08-62-0635	DATE	7-7
MATERIAL	DESCRIPTION	Ø 3/8 41L40 STL.	DRAWN BY	PAB
	PART NUMBER	36-66-4289	CHECKED BY	SK

UNLESS OTHERWISE SPECIFIED

FINISH SPECIFICATIONS
MACHINED SURFACES 125 MAX.
GROUND SURFACES 32 MAX.
DECIMAL TOL. XX ± .010 XXX ± .005 XXXX ± .0005
FRACTION TOL. ± .010
ANGLE TOL. ± 1°
REMOVE BURRS AND SHARP EDGES TO .015 MAX

DO NOT SCALE

G	
F	
E	
D	
C	
B	
A	
	CHANGE	ECO	DATE

INTER. PINION BLANK QUIZ

1. How long is the .330 diameter?

1. _____

2. How wide are the two grooves in the workpiece?

2. _____

3. What size chamfers are shown on the print?

3. _____

4. What is the high limit on the overall length of the workpiece?

4. _____

5. What scale is the print?

5. _____

6. Determine dimension [A] .

6. _____

7. What are the low limits for the groove diameters?

7. _____

8. What type of line is [B] ?

8. _____

9. What kind of material is used for the part?

9. _____

10. What type of line is [C] ?

10. _____

11. What is the part number?

11. _____

12. What is the tolerance on three place decimal dimensions?

12. _____

13. Who checked the print?

13. _____

14. What drawing number is the print?

14. _____

15. What size stock is used to make the part?

15. _____

Unit 11

NECKS, GROOVES, SLOTS, KEYWAYS, KEYSEATS, FLATS, BOSSES, AND PADS

After studying this unit, you will be able to:
☐ Determine the difference between a neck and a groove.
☐ Determine the sizes of necks and grooves.
☐ Identify common types of slots.
☐ Identify type and size of a keyway and keyseat.
☐ Distinguish between a boss and a pad.
☐ Identify a flat.

This unit groups together various topics that are important to interpreting elementary level industrial prints. Six of the eight topics share commonality and can be grouped under two headings: the fastening group and the assembly group. Each of the six topics (necks, grooves, slots, keyways, keyseats, and flats) is produced through machining operations. All six are used in the fastening and assembly of manufactured parts.

The remaining two topics, bosses and pads, deal with material allowance found on workpieces.

NECKS

NECKS (sometimes called ''grooves'') are recesses in the outside diameters of workpieces to allow mating objects to fit flush to each other. See Fig. 11-1. Necks are frequently used when it is necessary to have a threaded diameter assemble flush to a shoulder. See Fig. 11-2.

Usually, a radius is found between stepped diameters of a workpiece. However, when that radius interferes with the proper assembly of two objects, a necking operation is performed, Fig. 11-3.

Necks can be dimensioned by stating the width and depth of the recess, using a note, or by stating the width and diameter of the recess, Fig. 11-4. When the size of the neck is not important, no dimension is given.

GROOVES

ANNULAR (ring-like) GROOVES can be found on both inside and outside diameters of workpieces,

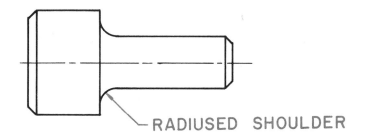
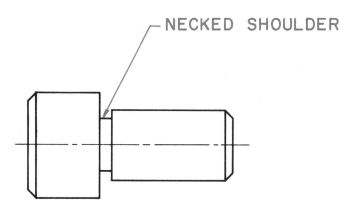

RADIUSED SHOULDER

NECKED SHOULDER

Fig. 11-1. Drawings show radiused and necked shoulders on machined parts.

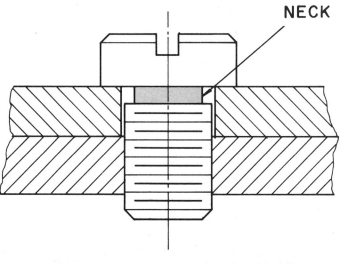

NECK

Fig. 11-2. Neck is used on a threaded fastener.

144

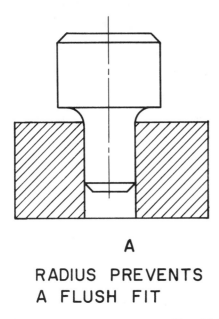

A

RADIUS PREVENTS
A FLUSH FIT

B

NECK ALLOWS FOR
A FLUSH FIT

Fig. 11-3. Using a neck for a flush fit.

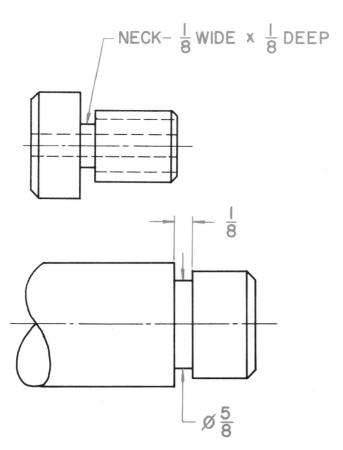

NECK— $\frac{1}{8}$ WIDE x $\frac{1}{8}$ DEEP

$\frac{1}{8}$

Ø $\frac{5}{8}$

Fig. 11-4. Two methods are shown for dimensioning necks.

Fig. 11-5. Grooves are frequently used for mounting fasteners such as snap rings and retaining rings. They are also used for installing seals such as O-rings and for passageways for lubricating oils.

Annular grooves may be dimensioned by stating the diameter and width or by providing the depth and width. See Fig. 11-6.

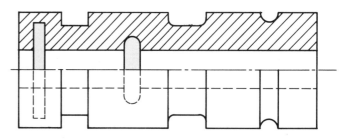

Fig. 11-5. Examples are given for various annular grooves.

The V-GROOVE is a familiar type of groove found on pulleys used in conjunction with V-belts. V-grooves are produced in many forms, which vary in angles, widths, and depth. See Fig. 11-7.

Some shafts have specially designed grooves as shown in Fig. 11-8. This requires the use of special ground (machined) tooling.

An INTERNAL GROOVE is sometimes called a "recess" or "undercut." Some EXTERNAL GROOVES on cylindrical workpieces may be called a "neck" when they occur at a shoulder.

SLOTS

Two principal types of SLOTS used on machines and other parts include the tee slot (T-slot) and the dovetail slot, Fig. 11-9. TEE SLOTS are used on machine tables for the purpose of fastening down devices such as vises, clamps, straps, fixtures, etc. T-BOLTS are used with the table to fasten down the holding devices and/or workpieces.

DOVETAIL SLOTS are used on machine tools as slides, creating an interlocking assembly between two machine parts to provide a reciprocating movement.

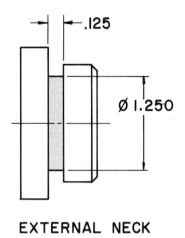

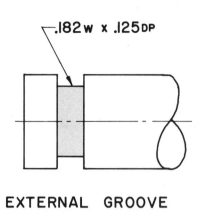

EXTERNAL NECK

EXTERNAL GROOVE

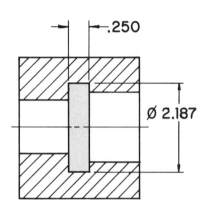

RECESS or INTERNAL GROOVE

Fig. 11-6. Drawings present methods of dimensioning external and internal grooves.

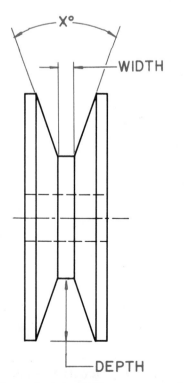

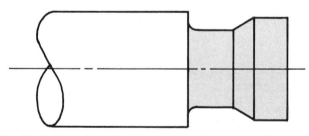

Fig. 11-8. Drawing shows an example of a special groove on a shaft end.

Common applications of dovetails include: cross-slide and compound rest of a lathe; table and saddle of a milling machine; tool-slide on an adjustable boring head.

KEYWAYS AND KEYSEATS

A KEYWAY is an internal groove machined in a hole along its length which provides a slot for inserting a key.

A KEYSEAT is an external groove cut into a shaft along its length to provide a seat for a key.

There are many types of keys used in keyways and keyseats. Common keys include the flat,

Fig. 11-7. Dimensioning requirements are given for a V-groove on a pulley.

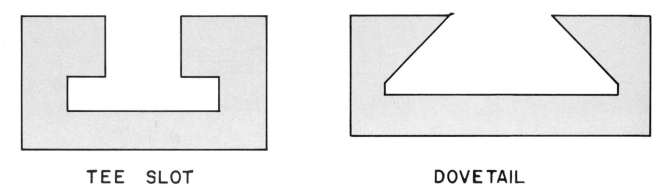

TEE SLOT   DOVETAIL

Fig. 11-9. Examples are given for a tee slot and a dovetail slot, used on machine tools.

square, and woodruff types. See Figs. 11-10, 11-11, and 11-12.

A key is used to secure a part to a shaft. Parts secured in this manner include gears, pulleys, collars (spacers), cranks, handles, handwheels, and cutting tools. A key provides a positive force and prevents the part from slipping on the shaft.

Various tolerances are used for keys, keyways, and keyseats. These specifications can be found in reference texts such as the *Machinery's Handbook* and the *American Machinists' Handbook*.

Keyways and keyseats for flat and square keys are dimensioned with leaders. The width dimension is given first, followed by the depth. The length is given by using a direct dimension on the print. See Fig. 11-13.

Woodruff keyseats, which are semicircular in shape, are dimensioned with leaders. The number of the woodruff keyseats and its location along the shaft are given. See Fig. 11-14. These keyseats are machined with special cutters called "woodruff cut-

ters." For positive identification, the number of the woodruff cutter corresponds identically with the number of the woodruff key.

The number of the woodruff key identifies its size. The last two digits of the number give the diameter of the key in eighths of an inch. The digits preceding the last two digits give the width of the key in thirty-seconds of an inch.

For example: 807 key

The number means the key is 8/32 in. × 7/8 in. or 1/4 in. wide × 7/8 in. diameter.

For precision fits such as required for interchangeable assemblies, keyway and keyseat dimensions are given with limit dimensions, Fig. 11-15, to provide proper fits.

FLATS

A FLAT is a depression found on a shaft or shank which provides a seat for a setscrew. The setscrew is used with another part to hold that object in place

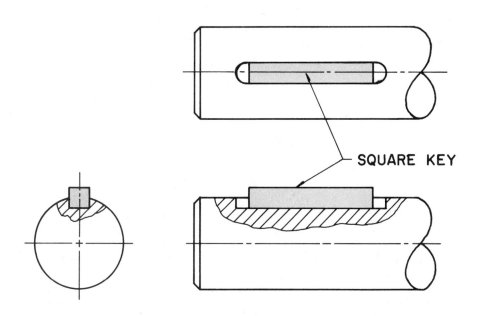

SQUARE KEY

Fig. 11-10. Three views show how a square key fits into a keyseat in a shaft.

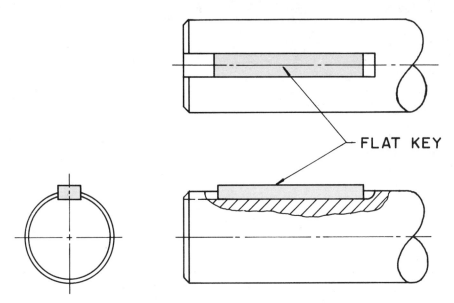

Fig. 11-11. Three views illustrate fit of a flat key in a keyseat.

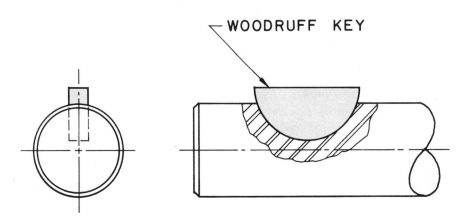

Fig. 11-12. Two views picture a woodruff key in a keyseat in a shaft.

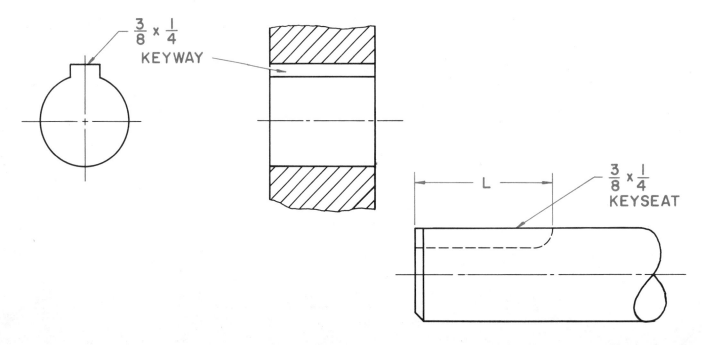

Fig. 11-13. Dimensioning requirements are shown: A—For a keyway (flat key). B—For a keyseat (flat key).

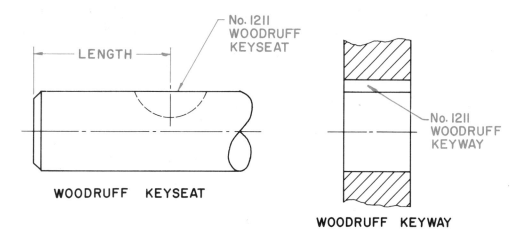

Fig. 11-14. Woodruff key number (1211) identifies its size: 12/32 x 1 1/8 or 3/8 x 1 3/8 (see text for explanation).

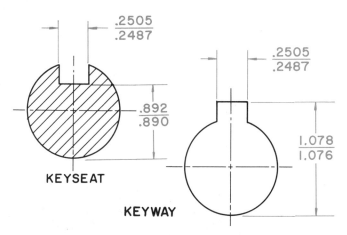

Fig. 11-15. Precision fits of keys, keyseats, and keyways require limit dimensioning on drawing.

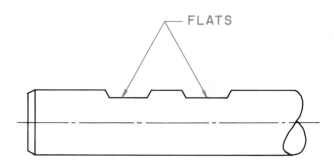

Fig. 11-16. Drawing depicts flats on a tool shank.

on the shaft or shank. Many straight tool shanks have flats on them, Fig. 11-16, to secure the tool to a tool holder.

BOSSES AND PADS

A BOSS is a cylindrical (round) raised surface found on a casting or forging which provides additional material to the part.

A PAD is a raised surface shaped other than round found on a casting to provide additional material to the workpiece.

Both bosses and pads are machined to provide smooth surfaces for mating parts. Bosses usually have holes machined in or through them, while pads are commonly found with a slot milled through them. See Fig. 11-17.

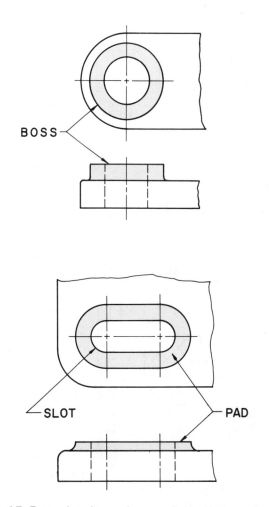

Fig. 11-17. Examples show a boss and a pad in two views.

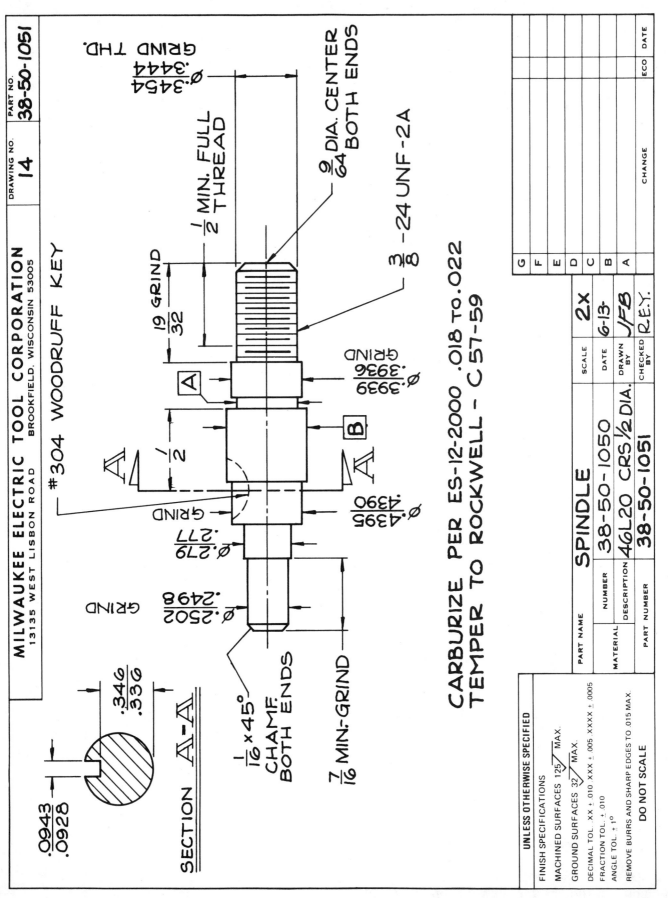

Fig. 11-18. Spindle.

DIRECTIONS—QUIZ QUESTIONS

1. The industrial prints in this section will test your print reading ability.
2. Study the views, dimensions, title block, and notes in Figs. 11-18 through 11-25.
3. Read the quiz questions, refer to the print, and write your answers in the blanks provided.

SPINDLE QUIZ

1. What size is the thread?

1. _____

2. What tolerance is given for the 1/4 diameter?

2. _____

3. How wide is the keyseat?

3. _____

4. Which diameter has the smallest tolerance?

4. _____

5. What is the approximate size of dimension [A] ?

5. _____

6. What size is the woodruff key?

6. _____

7. What size is the outside diameter of the thread?

7. _____

8. What scale is the print?

8. _____

9. What diameters are ground?

9. _____

10. What is the approximate depth of the keyseat?

10. _____

11. What is dimension [B] ?

11. _____

12. Why are some length dimensions missing on the print?

12. _____

13. What dimension shows the location of the keyseat?

13. _____

14. What does ES mean?

14. _____

15. How long is the thread?

15. _____

16. To what depth is the part hardened?

16. _____

17. State what finish is required for ground surfaces.

17. _____

18. What thread series is specified on the print?

18. _____

19. What material is the part?

19. _____

20. What does 2A mean on the print?

20. _____

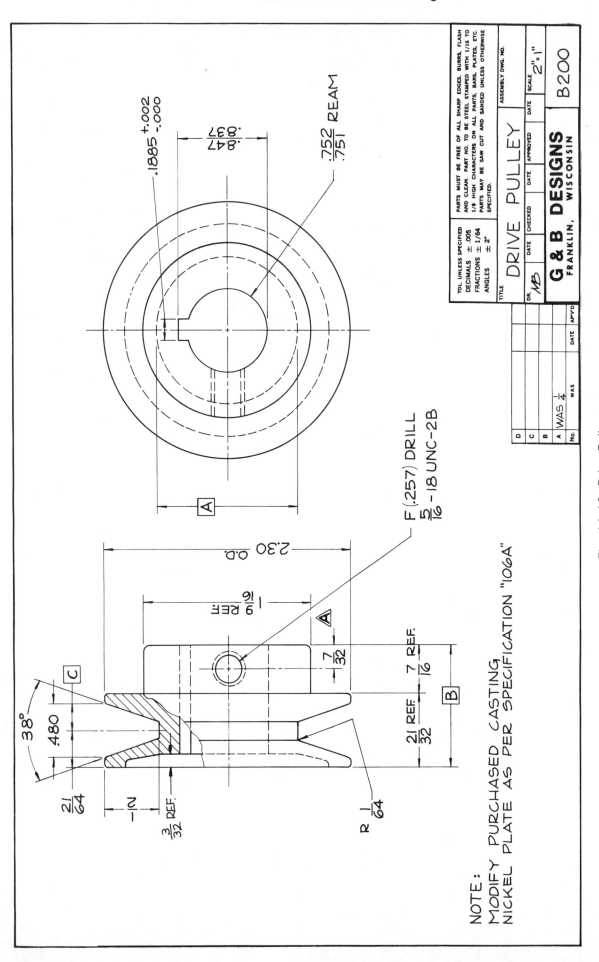

.1885 $^{+.002}_{-.000}$

.847
.837

.752 REAM
.751

F (.257) DRILL
$\frac{5}{16}$ - 18 UNC-2B

2.30 O.D.

1 $\frac{9}{16}$ REF.

$\frac{7}{32}$

38°

.480

$\frac{21}{64}$

2

$\frac{3}{32}$ REF.

$\frac{21}{32}$ REF.

$\frac{7}{16}$ REF.

R $\frac{1}{64}$

A

B

C

NOTE:
MODIFY PURCHASED CASTING
NICKEL PLATE AS PER SPECIFICATION "106A"

TOL. UNLESS SPECIFIED	PARTS MUST BE FREE OF ALL SHARP EDGES, BURRS, FLASH AND CLEAN. PART NO. TO BE STEEL STAMPED WITH 1/16 TO 1/8 HIGH CHARACTERS ON ALL PARTS, BARS, PLATES, ETC. PARTS MAY BE SAW CUT AND SANDED UNLESS OTHERWISE SPECIFIED.
DECIMALS ± .005	
FRACTIONS ± 1/64	
ANGLES ± 2°	

TITLE
DRIVE PULLEY

DR. *MB*

G & B DESIGNS
FRANKLIN, WISCONSIN

DATE | CHECKED | DATE | APPROVED | DATE

ASSEMBLY DWG. NO.

SCALE 2"=1"

B200

D					
C					
B					
A	WAS $\frac{1}{4}$		WAS		
No.			DATE		APV'D

Fig. 11-19. Drive Pulley.

DRIVE PULLEY QUIZ

1. How wide is the drive pulley groove at the outside?   1. _____

2. How wide is the hub of the pulley?   2. _____

3. How wide is the pulley keyway?   3. _____

4. How deep is the pulley groove?   4. _____

5. What is the maximum outside diameter of the pulley?   5. _____

6. What is dimension A ?   6. _____

7. How long is the threaded hole?   7. _____

8. What is the approximate length of the keyway?   8. _____

9. What is dimension B ?   9. _____

10. What kind of material is used for the pulley?   10. _____

11. What angle is the groove?   11. _____

12. What diameter is the hub of the pulley?   12. _____

13. What is the low limit of the reamed hole?   13. _____

14. What does the large hidden circle represent?   14. _____

15. How far is the tapped hole from the end of the hub?   15. _____

16. What was revision A ?   16. _____

17. What finish is required on the pulley?   17. _____

18. What does REF. mean?   18. _____

19. What is the maximum dimension from the center of the pulley to the bottom of the keyway?   19. _____

20. What is dimension C ?   20 _____

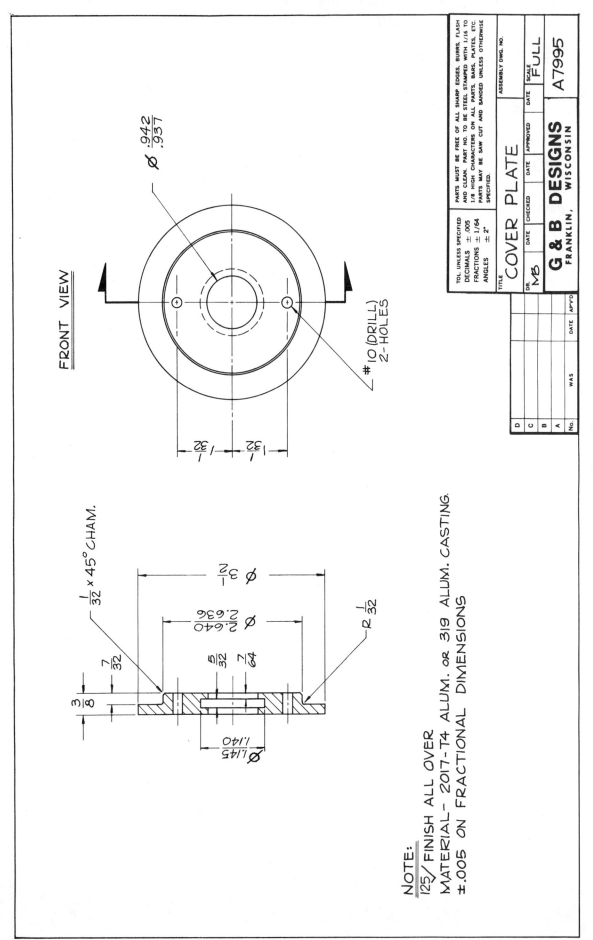

FRONT VIEW

$\emptyset \frac{.942}{.937}$

#10 (DRILL)
2-HOLES

$1\frac{1}{32}$ $1\frac{1}{32}$

$\frac{1}{32} \times 45° \text{ CHAM.}$

$\emptyset 3\frac{1}{2}$

$\emptyset \frac{2.640}{2.636}$

$R\frac{1}{32}$

$\frac{7}{32}$

$\frac{3}{8}$

$\frac{5}{32}$

$\frac{7}{64}$

$\emptyset \frac{1.145}{1.140}$

NOTE:
$\frac{125}{}$ FINISH ALL OVER
MATERIAL - 2017-T4 ALUM. OR 319 ALUM. CASTING.
±.005 ON FRACTIONAL DIMENSIONS

TOL. UNLESS SPECIFIED
DECIMALS ± .005
FRACTIONS ± 1/64
ANGLES ± 2°

PARTS MUST BE FREE OF ALL SHARP EDGES, BURRS, FLASH AND CLEAN. PART NO. TO BE STEEL STAMPED WITH 1/16 TO 1/8 HIGH CHARACTERS ON ALL PARTS, BARS, PLATES, ETC. PARTS MAY BE SAW CUT AND SANDED UNLESS OTHERWISE SPECIFIED.

ASSEMBLY DWG. NO.

TITLE
COVER PLATE

DR. MB | DATE | CHECKED | DATE | APPROVED | DATE | SCALE FULL

G & B DESIGNS
FRANKLIN, WISCONSIN

A7995

No.	WAS	DATE	APV'D
D			
C			
B			
A			

Fig. 11-20. Cover Plate.

COVER PLATE QUIZ

1. What type of sectional view is shown on the print?
2. What does the dashed line circle in the front view represent?
3. How many holes are in the part?
4. How long is the 3 1/2 diameter?
5. What is the decimal equivalent of a #10 drill?
6. Determine the high limit on the largest diameter.
7. How far apart are the two #10 holes from each other?
8. What finish is required on the part?
9. What is the maximum diameter allowed on the groove?
10. How wide is the groove?
11. What is the overall length (thickness) of the part?
12. How many surfaces are shown in the front view?
13. What does the number 319 refer to on the print?
14. Is the groove in the center (middle) of the part?
15. What size is the large hole on the print?

1. _____
2. _____
3. _____
4. _____
5. _____
6. _____
7. _____
8. _____
9. _____
10. _____
11. _____
12. _____
13. _____
14. _____
15. _____

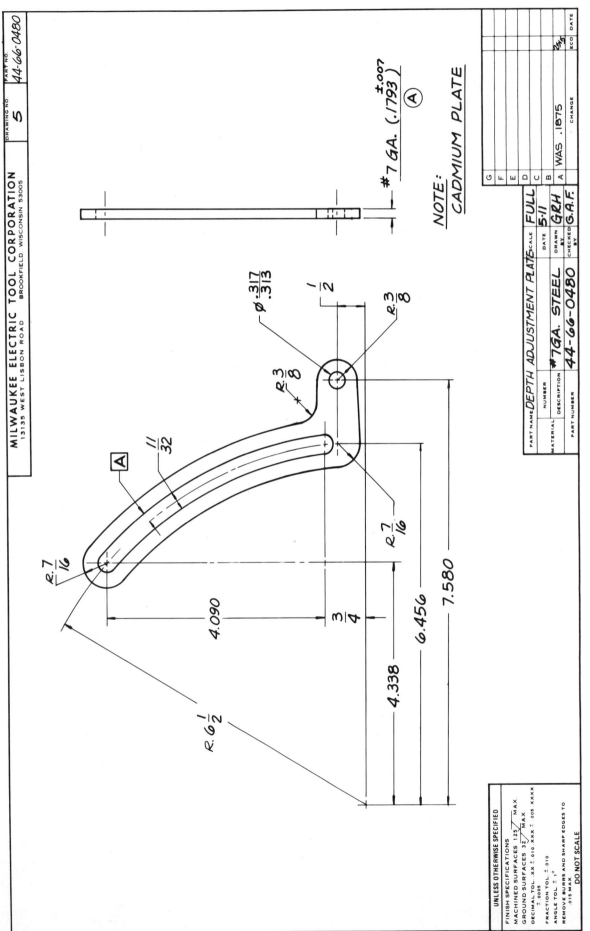

Fig. 11-21. Depth Adjustment Plate.

DEPTH ADJUSTMENT PLATE QUIZ

1. List all radii found on the print.   1. _____

2. How thick is the part?   2. _____

3. Give the two dimensions describing the location of the hole.   3. _____

4. How wide is the slot?   4. _____

5. On what radius is the slot located?   5. _____

6. What is the actual height of the part?   6. _____

7. What type of finish does the part receive?   7. _____

8. What thickness was the part originally?   8. _____

9. What is the part number?   9. _____

10. What is dimension [A] ?   10. _____

11. What is the width of the part?   11. _____

12. What type of decimal dimension is used on the .1793 dimension?   12. _____

13. What is the scale of the print?   13. _____

14. How far is the horizontal centerline of the hole from the upper horizontal centerline of the slot?   14. _____

15. How far apart are the horizontal centerlines of the slot?   15. _____

16. What is the maximum thickness allowed for the part?   16. _____

17. At what distance is the horizontal centerline of the hole from the lower horizontal centerline of the slot?   17. _____

18. How far apart are the vertical centerlines of the slot?   18. _____

19. How many centerlines are shown on the print?   19. _____

20. What is the low limit for the 7.580 dimension?   20. _____

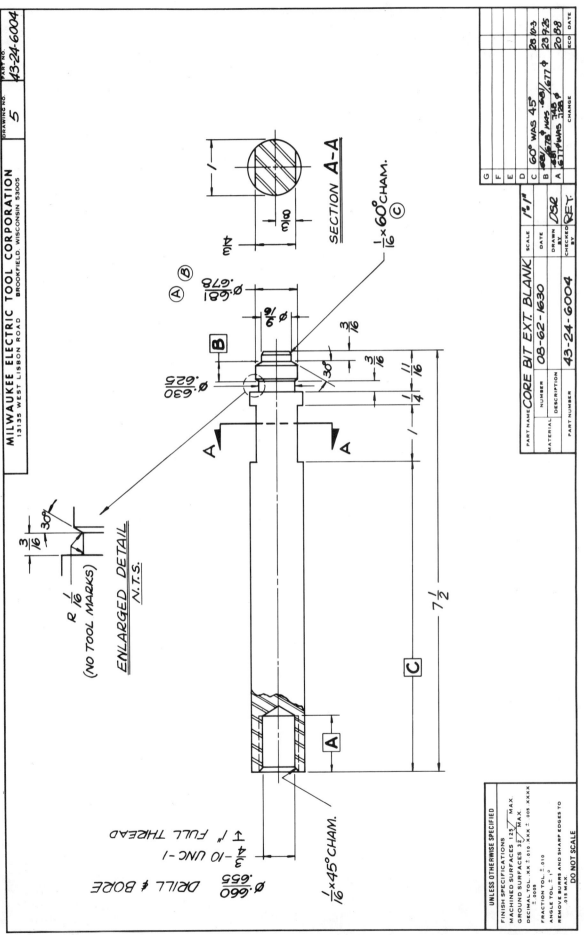

Fig. 11-22. Core Bit Ext. Blank.

CORE BIT EXT. BLANK QUIZ

1. How long are the two flats shown on the print?

2. What is the maximum allowable total length of the part?

3. What size is the groove diameter?

4. What size chamfer is used on the thread?

5. How wide is the groove?

6. Is there a fit designated for the thread? If yes, what is it?

7. What is dimension A ?

8. What is the low limit for the smallest diameter shown on the print?

9. What note refers to the external groove on the print?

10. Determine dimension B .

11. What is the distance between the two flats?

12. What does N.T.S. mean?

13. What is the outside diameter of the part?

14. What is revision C ?

15. What kind of sectional view is shown in the front view?

16. What kind of material is used for the part?

17. What engineering change order was used for Revision B?

18. What kind of shoulder is found between the 9/16 diameter and the .681/.678 diameter?

19. Determine dimension C .

20. How many centerlines are found on the print?

1. _____
2. _____
3. _____
4. _____
5. _____
6. _____
7. _____
8. _____
9. _____
10. _____
11. _____
12. _____
13. _____
14. _____
15. _____
16. _____
17. _____
18. _____
19. _____
20. _____

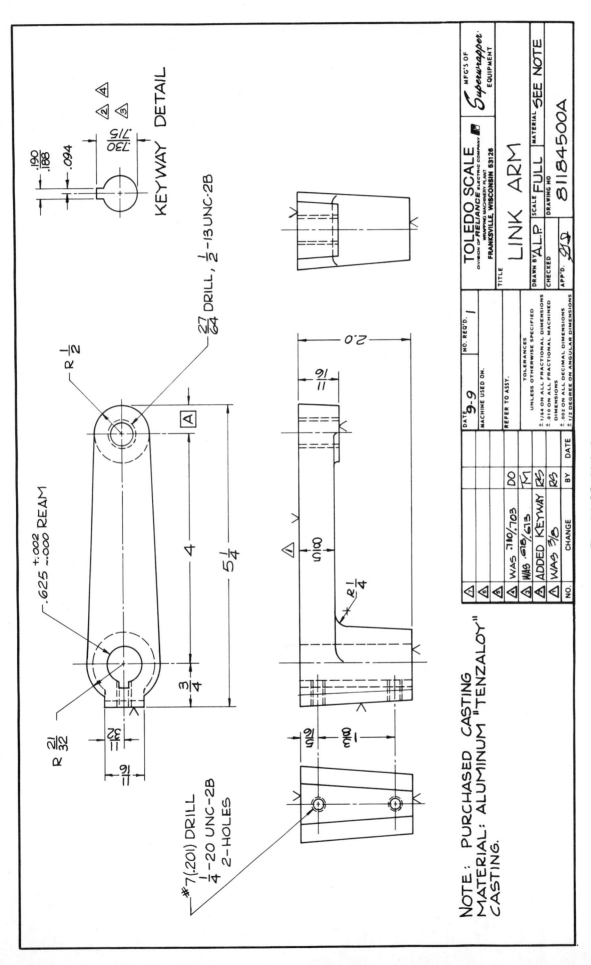

Fig. 11-23. Link Arm.

LINK ARM QUIZ

1. What is the minimum total height of the part?

2. What is the total length of the part?

3. What is revision △1 ?

4. How long (deep) is the 1/2-13 UNC thread?

5. What diameter is the boss on the part?

6. How much clearance is there between the keyway and a 3/16 key?

7. How far apart are the 1/4 tapped holes?

8. Was the keyway depth increased or decreased from the last revision?

9. How far is the reamed hole from the 1/2-13 UNC threaded hole?

10. Is the right side view necessary?

11. How much does the part taper in its length? Give answer in TPI.

12. Determine dimension A .

13. How much was the thickness of the link arm base increased?

14. What kind of material is "TENZALOY"?

15. What is the approximate depth of the keyway?

16. How many surfaces are finished?

17. What is the lower limit of the reamed hole?

18. What size tap drills are used on the print?

19. How far does the top of the boss extend above the top of the base?

20. Where is Toledo Scale located?

1. _____

2. _____

3. _____

4. _____

5. _____

6. _____

7. _____

8. _____

9. _____

10. _____

11. _____

12. _____

13. _____

14. _____

15. _____

16. _____

17. _____

18. _____

19. _____

20. _____

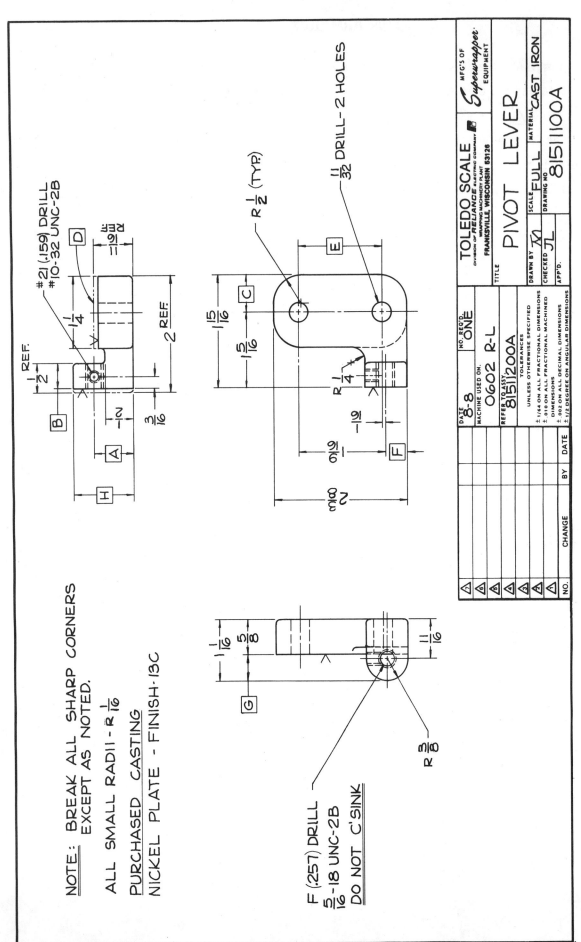

Fig. 11-24. Pivot Lever.

PIVOT LEVER QUIZ

1. Determine dimension $\boxed{A}$.

2. How many workpieces are required?

3. What size are the small radii on the print?

4. Determine dimension $\boxed{B}$.

5. How far are the two drilled holes from the centerline of the small tapped hole?

6. What is dimension $\boxed{C}$?

7. What finish is required on the casting?

8. What kind of line is $\boxed{D}$?

9. What do the REF. dimensions represent?

10. Determine dimension $\boxed{E}$.

11. What size is the fillet on the casting?

12. Determine dimension $\boxed{F}$.

13. What does the "B" signify on the thread designation 5/16—18 UNC—2B?

14. What machine is this part used on?

15. What views are shown on this print?

16. Determine dimension $\boxed{G}$.

17. What is the width of the large machined surface?

18. What tolerance is used on fractional dimensions?

19. Determine dimension $\boxed{H}$.

20. Which view shows the shape of the small machined surface?

1. _____

2. _____

3. _____

4. _____

5. _____

6. _____

7. _____

8. _____

9. _____

10. _____

11. _____

12. _____

13. _____

14. _____

15. _____

16. _____

17. _____

18. _____

19. _____

20. _____

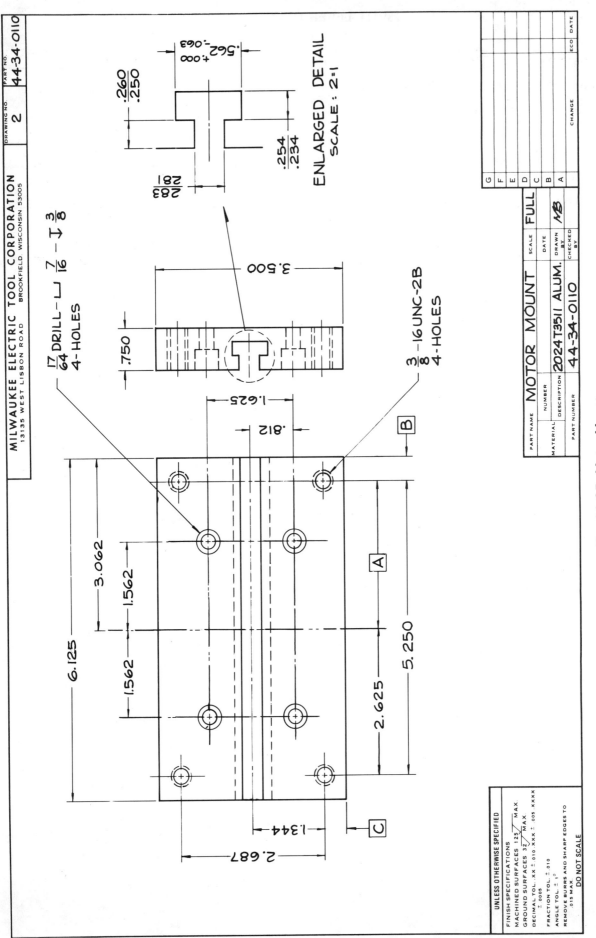

Fig. 11-25. Motor Mount.

Necks, Grooves, Slots, Keyways, Keyseats, Flats, Bosses, and Pads

MOTOR MOUNT QUIZ

1. How wide is the opening of the T-slot?

1. _____

2. What is the total length of the workpiece?

2. _____

3. How many holes are tapped?

3. _____

4. Determine dimension [A] .

4. _____

5. How thick is the workpiece?

5. _____

6. How far apart horizontally are the two tapped holes on the left from the two counterbore holes on the left?

6. _____

7. What is the minimum total depth of the T-slot?

7. _____

8. What is the maximum total depth of the T-slot?

8. _____

9. Determine dimension [B] .

9. _____

10. What size are the counterbores?

10. _____

11. How far apart are the two counterbore holes on the left from the two counterbore holes on the right?

11. _____

12. What is the upper limit dimension for the width at the bottom of the T-slot?

12. _____

13. How far are the 3/8-16 tapped holes from the horizontal centerline of the workpiece?

13. _____

14. Determine dimension [C] .

14. _____

15. How deep are the counterbores?

15. _____

16. How far apart are the two upper counterbore holes from the two lower tapped holes?

16. _____

17. What is the lower limit dimension for the width at the bottom of the T-slot?

17. _____

18. What is the upper limit dimension on the width of the workpiece?

18. _____

19. What size are the tapped holes?

19. _____

20. What is the maximum thickness from the bottom of the T-slot to the bottom of the workpiece?

20. _____

165

Unit 12

GEOMETRIC DIMENSIONING AND TOLERANCING

After studying this unit, you will be able to:
☐ Identify various geometric symbols.
☐ Identify datum surfaces.
☐ Define various terms relating to geometric tolerancing.
☐ Interpret the meaning of geometric symbols on a print.

Advancements in technology have brought about greater control in the accuracy of machined parts. Still, it is almost impossible to manufacture a perfect part, so it becomes necessary to define the amount of variation permitted to a specific form of a part. These factors have led to the use of a drafting system commonly known as ''geometric dimensioning and tolerancing.''

The methods of applying geometric dimensioning and tolerancing to conventional dimensioning is through the use of geometric symbols that have been recommended by the American National Standards Institute (ANSI).

This unit, then, is a brief introduction to geometric tolerancing. Information contained in the unit is a partial, summary version of U.S. dimensioning and tolerancing practices. For more complete and detailed information, refer to the publication: American National Standard Y14.5M—1982, DIMENSIONING AND TOLERANCING.

TERMS

Several TERMS appearing in this unit need to be defined prior to their use:

FEATURE: Universal term applied to an actual portion of a part, such as a surface, hole, thread, or groove.

DATUM: Exact point, axis, or plane serving as the origin from which location or geometric characteristics of features of a part are derived.

DATUM FEATURE: Actual feature of a part used as a datum.

BASIC DIMENSION: A numerical value stating the exact size, contours, orientation, or position of a feature or datum.

TOLERANCE: Total amount a dimension is allowed to vary. The difference between maximum and minimum limits.

LIMITS: The maximum and minimum size allowed on a dimension.

Various other new terms will be defined as they appear within this unit.

APPLICATION

GEOMETRIC TOLERANCING is specifying the allowable variation permitted on exact form or true position on part features.

Geometric tolerancing is applied to five areas of concern: Form, Profile, Orientation, Locational, and Runout Tolerances.

FORM tolerances control the form or shape of various geometric figures. Form tolerances control straightness, flatness, circularity (roundness), and cylindricity. Form tolerances are not related to datums.

PROFILE tolerances are used to control form, or combination of size, form, and orientation. They specify a constant boundary along the true profile within which all points or elements of the surface must lie. Profile tolerances include line profile and surface profile.

ORIENTATION tolerances control angularity, perpendicularity, and parallelism.

LOCATIONAL tolerances define the allowable variation of a feature from the exact or true position shown on the drawing.

RUNOUT tolerances control the relationships of one or more features of a part to its axis.

STANDARD SYMBOLS

The accompanying chart, Fig. 12-1, shows the various symbols used in describing the characteristics found in Form, Profile, Orientation, Locational,

	TYPE OF TOLERANCE	CHARACTERISTIC	SYMBOL	SEE:
FOR INDIVIDUAL FEATURES	FORM	STRAIGHTNESS	—	6.4.1
		FLATNESS	▱	6.4.2
		CIRCULARITY (ROUNDNESS)	○	6.4.3
		CYLINDRICITY	⌀	6.4.4
FOR INDIVIDUAL OR RELATED FEATURES	PROFILE	PROFILE OF A LINE	⌒	6.5.2 (b)
		PROFILE OF A SURFACE	⌓	6.5.2 (a)
FOR RELATED FEATURES	ORIENTATION	ANGULARITY	∠	6.6.2
		PERPENDICULARITY	⊥	6.6.4
		PARALLELISM	∥	6.6.3
	LOCATION	POSITION	⊕	5.2
		CONCENTRICITY	◎	5.11.3
	RUNOUT	CIRCULAR RUNOUT	↗ *	6.7.2.1
		TOTAL RUNOUT	↗↗ *	6.7.2.2

*Arrowhead(s) may be filled in.

Fig. 12-1. Geometric characteristic symbols. (The American Society of Mechanical Engineers)

and Runout tolerances. Other standard symbols adopted by ANSI Y14.5M—1982 and used in this unit include: ϕ, R, ⌴, ⌵, ⊤, and □ .

The symbol ϕ means DIAMETER. It is used instead of the word "diameter" before the actual size. It appears as ϕ1.00, meaning one inch diameter.

The symbol R means RADIUS. It is used instead of the word "radius" before the actual size. It would appear as R 1.5, meaning 1 1/2 inch radius.

The symbol ⌴ represents a SPOTFACE or COUNTERBORE. It appears in front of the dimensions of a spotface or counterbore. See Fig. 12-2.

The symbol ⌵ represents a COUNTERSINK. It appears before the dimensions of a countersink as in Fig. 12-3.

The symbol ⊤ shown in Fig. 12-4 represents the word DEPTH. It precedes that dimension and refers to the depth of a feature, such as a hole.

The symbol □ indicates that a part feature is SQUARE. This symbol precedes the dimension denoting the size of the square. See Fig. 12-5.

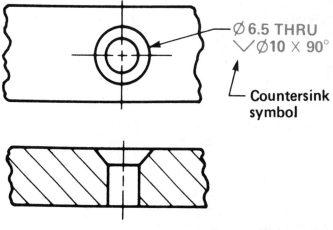

Ø 6.5 THRU
⌵ Ø10 × 90°

Countersink
symbol

Fig. 12-3. Countersink symbol. (ASME)

GEOMETRIC TOLERANCE SYMBOLS

GEOMETRIC TOLERANCE SYMBOLS are enclosed in a rectangular box containing the allowed

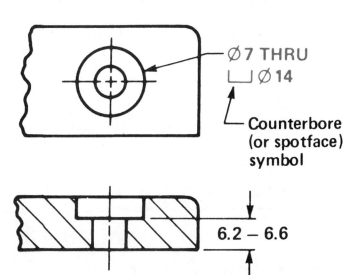

Ø7 THRU
⌴ Ø 14

Counterbore
(or spotface)
symbol

6.2 – 6.6

Fig. 12-2. Counterbore or spotface symbol. (ASME)

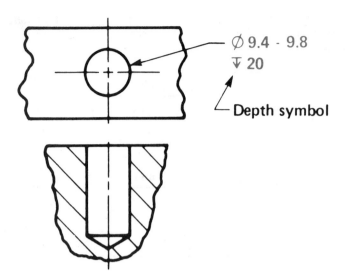

Fig. 12-4. Depth symbol. (ASME)

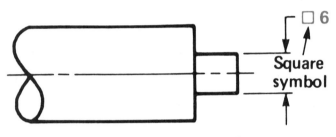

Fig. 12-5. Square symbol. (ASME)

tolerance at datum with a leader pointing to the feature to which it applies, Fig. 12-6.

DATUM FEATURE SYMBOL

The DATUM SYMBOL shown in Fig. 12-7 consists of a rectangular frame containing the datum reference letter. The letter is preceded and followed by a dash (—).

FEATURE CONTROL SYMBOL

A locational or form tolerance is characterized by using a FEATURE CONTROL SYMBOL. It consists of a rectangular frame containing the geometric character, followed by the allowable tolerance. See Fig. 12-8.

FEATURE CONTROL SYMBOL USING DATUM REFERENCES

When a locational or form tolerance must be related to a reference (datum), this relationship is stated in the feature control symbol. The datum reference letter is placed between the geometric symbol and the tolerance, as shown in Fig. 12-9.

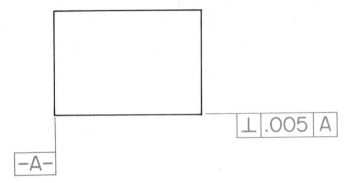

Fig. 12-6. Application of datum feature symbol and feature control frame.

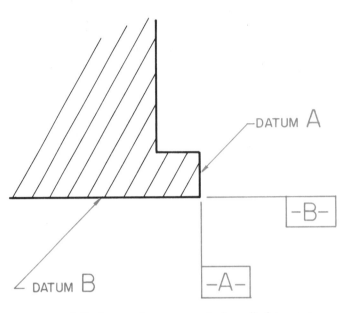

Fig. 12-7. Datum feature symbol applied to part.

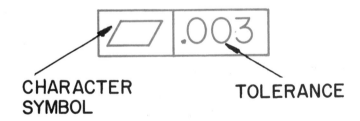

Fig. 12-8. Feature control frame.

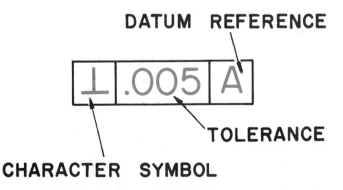

Fig. 12-9. Parts of feature control frame.

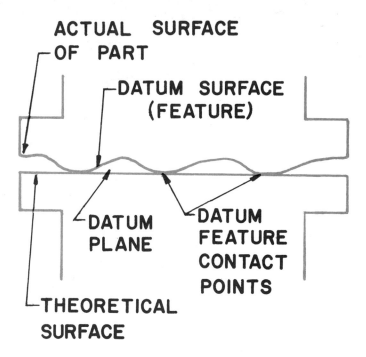

Fig. 12-10. Actual surface of part contacting theoretical (datum) surface.

DATUMS

A DATUM is defined as being a plane, point, line, or axis of a part. A datum is used as a reference base or exact position from which form or locational tolerances are checked.

DATUM PLANE/DATUM SURFACE

DATUM PLANES are theoretical exact reference bases.

DATUM SURFACES and FEATURES are actual surfaces or features of an object used to establish datums which include any surface and feature irregularities. See Fig. 12-10.

Datum surfaces can be controlled through the use of various symbols. For example, to control flatness for a datum surface, the symbol ⊿ is used, Fig. 12-11. Also note that the difference in height between the high points and low points of datum surface B must be within a tolerance of .005 inches. This is the surface flatness tolerance of the part.

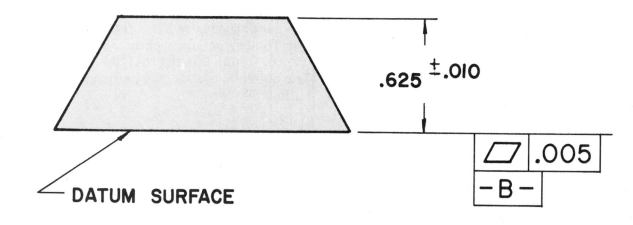

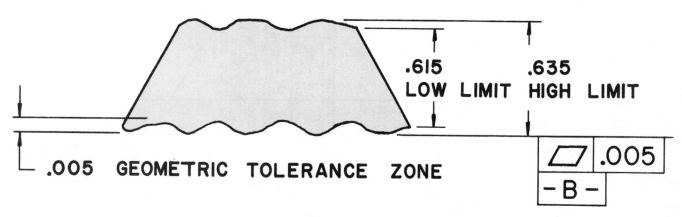

Fig. 12-11. Specifying flatness of a surface.

Besides the flatness tolerance required for surface B, a size tolerance of ± .010 inch must be held between datum surface B and another surface. The size between the high points of datum surface B and the high points of the other surface specified must not exceed the high limit of .635 inches. Also, the size between the low points of surface B and the low points of the other surface specified must not be less than the low limit of .615 inches.

In summary, when machining both surface B and the other surface, two concerns must be taken into account: datum surface B must be held to a flatness tolerance of .005 inch and still be within a size tolerance of ± .010 inch with another surface.

DATUM POINT

An entire surface does not always need to be machined. This often occurs on a casting or forging. When it does happen, only specific points that are to be machined are identified as datums.

DATUM POINTS are identified with the symbol ⊕. In this symbol, the A identifies the datum and the 2 identifies the point.

All datum points are located with basic dimensions. See Figs. 12-12 and 12-13.

Datum points can be controlled on a surface by

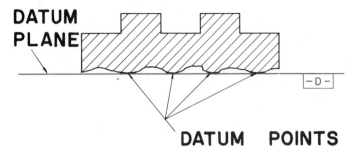

DATUM PLANE / DATUM POINTS

Fig. 12-13. Specific points identifying a datum.

using a ⌒ symbol which specifies profile.

In Fig. 12-14, the symbol ⌒ states that surface B, which is a contoured surface, must fall within a tolerance zone which is .010 inch wide. This means that the profile of surface B must not vary a total amount of .010 inch from high points to low points, and that surface B must be within a .010 inch wide boundary zone.

DATUM AXIS

A centerline is used as a datum feature on cylindrical and symmetrical parts. See Figs. 12-15 and 12-16. Datum B is the DATUM AXIS of the part. The datum axis shown is perpendicular to datum surface A.

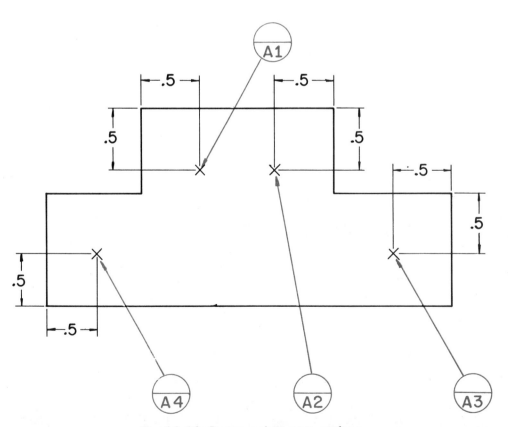

Fig. 12-12. Datum points on a surface.

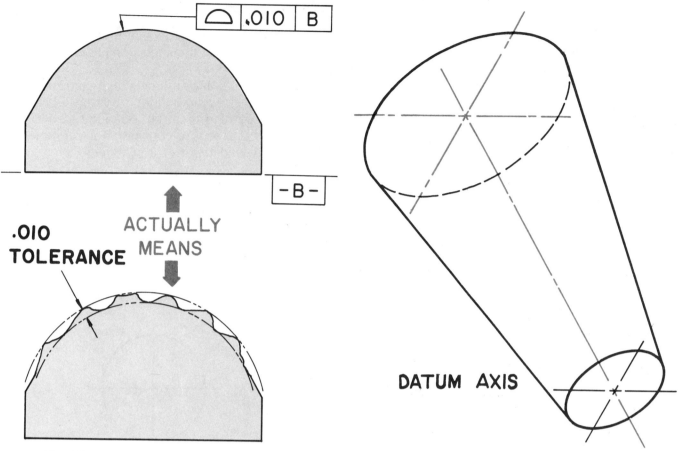

Fig. 12-14. Specifying profile of a surface.

Fig. 12-15. Datum axis on a cylindrical part.

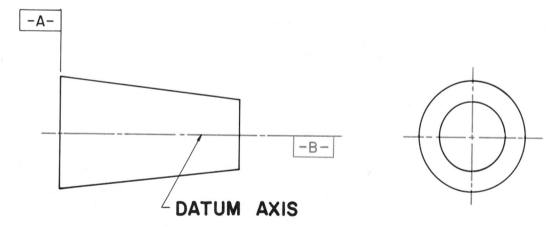

Fig. 12-16. Centerline as a datum axis.

GEOMETRIC SYMBOLS AND THEIR MEANING

Geometric symbols have been established for form tolerances, profile tolerances, orientation tolerances, locational tolerances, and runout tolerances.

FORM TOLERANCES

Form tolerance symbols depict roundness, cylindricity, straightness, and flatness.

Roundness ⌀

ROUNDNESS applies to a surface of revolution making all points of a surface intersected by a cutting plane perpendicular to a common axis and equidistant (same distance) from axis, Fig. 12-17.

Roundness tolerance states the maximum variation allowed between maximum and minimum radii, Fig. 12-18. The .005 inch wide tolerance zone for roundness is between two concentric circles. The actual round surface of the cylinder must lie within the .005 inch wide tolerance zone.

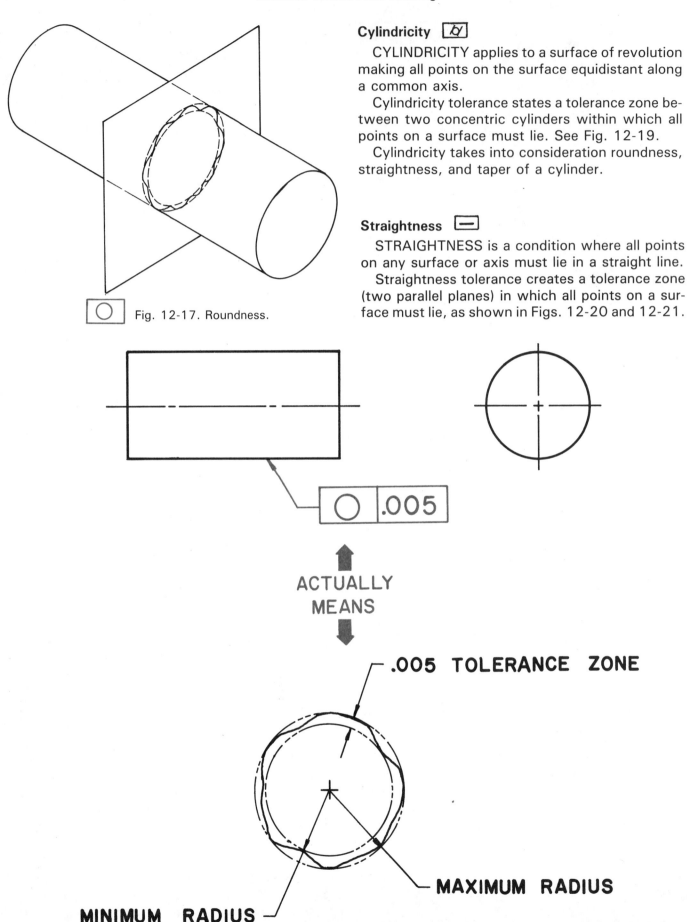

Cylindricity ⬡

CYLINDRICITY applies to a surface of revolution making all points on the surface equidistant along a common axis.

Cylindricity tolerance states a tolerance zone between two concentric cylinders within which all points on a surface must lie. See Fig. 12-19.

Cylindricity takes into consideration roundness, straightness, and taper of a cylinder.

Straightness ▭

STRAIGHTNESS is a condition where all points on any surface or axis must lie in a straight line.

Straightness tolerance creates a tolerance zone (two parallel planes) in which all points on a surface must lie, as shown in Figs. 12-20 and 12-21.

Fig. 12-17. Roundness.

ACTUALLY MEANS

.005 TOLERANCE ZONE

MAXIMUM RADIUS

MINIMUM RADIUS

Fig. 12-18. Roundness tolerance zone.

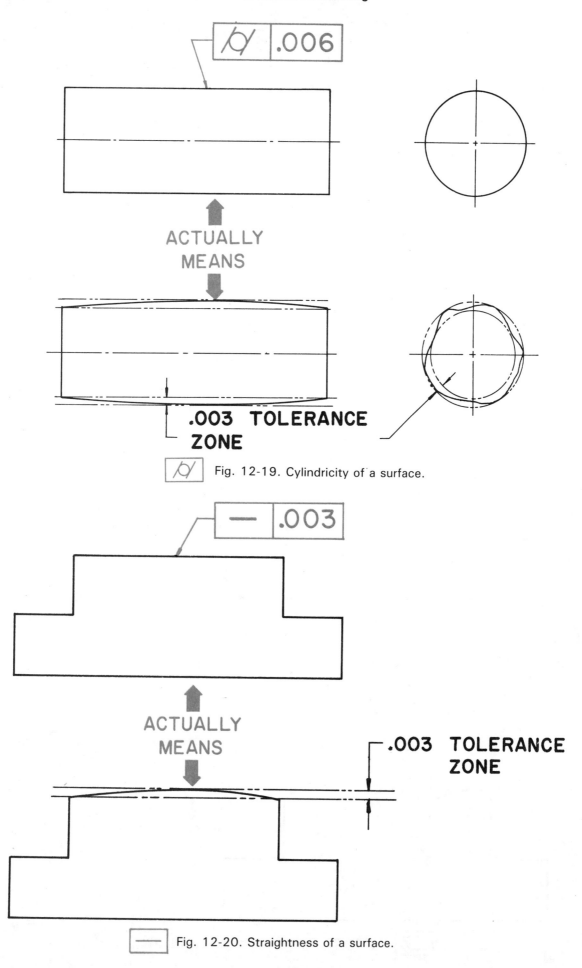

Fig. 12-19. Cylindricity of a surface.

Fig. 12-20. Straightness of a surface.

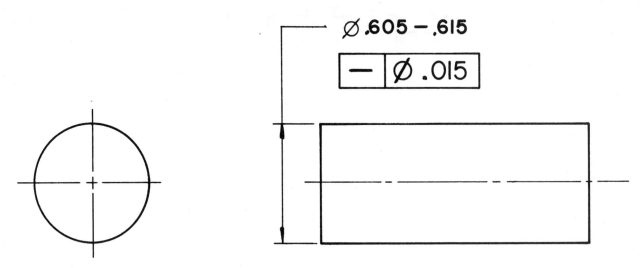

Fig. 12-21. Specifying axis straightness.

Flatness ▱

FLATNESS requires all points on a surface to lie in one plane.

Flatness tolerance creates a tolerance zone formed by two parallel planes between which the entire surface must lie. See Fig. 12-22.

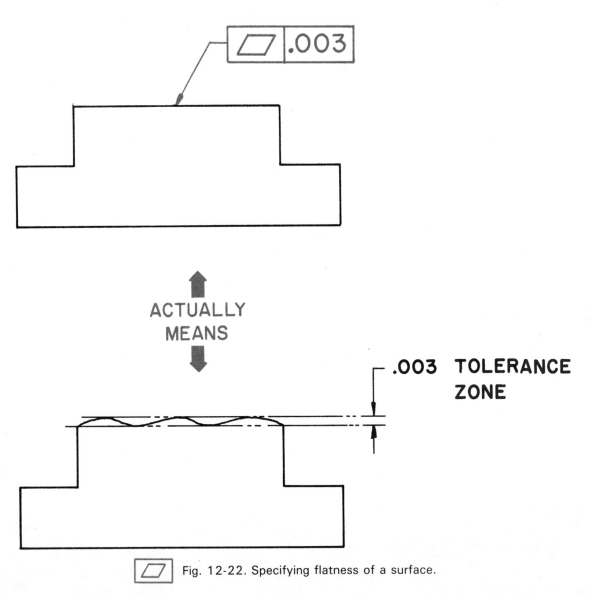

Fig. 12-22. Specifying flatness of a surface.

PROFILE TOLERANCES

Profile tolerance symbols include profile of a line and profile of a surface.

Profile (line)

PROFILE OF A LINE states all points on a line must lie within the specified tolerance zone of the true profile. Profile of a line is used when the entire surface control is not required. It is also used on shapes with varying cross-sectional dimensions along their length or width, as shown in Fig. 12-23.

Profile (surface)

PROFILE OF A SURFACE states all points along an entire surface must lie within the specified tolerance zone of the true profile. See Fig. 12-24.

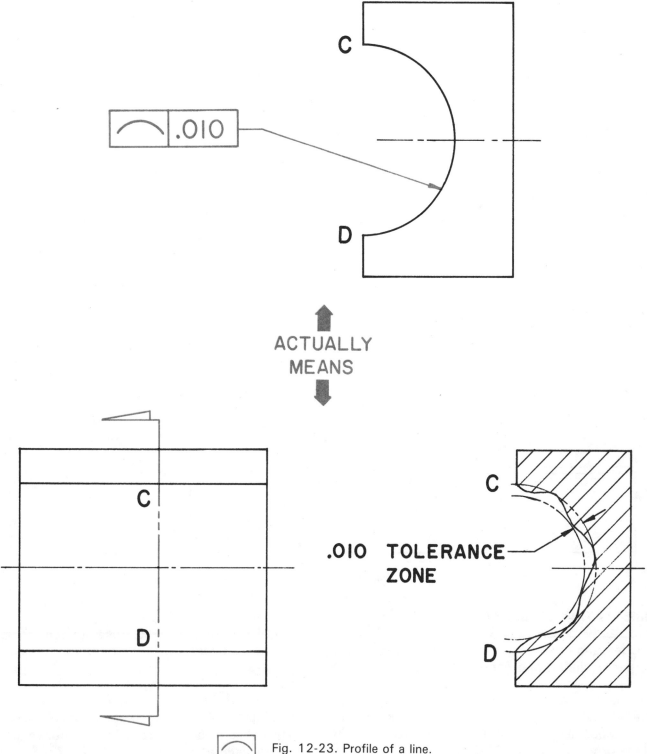

Fig. 12-23. Profile of a line.

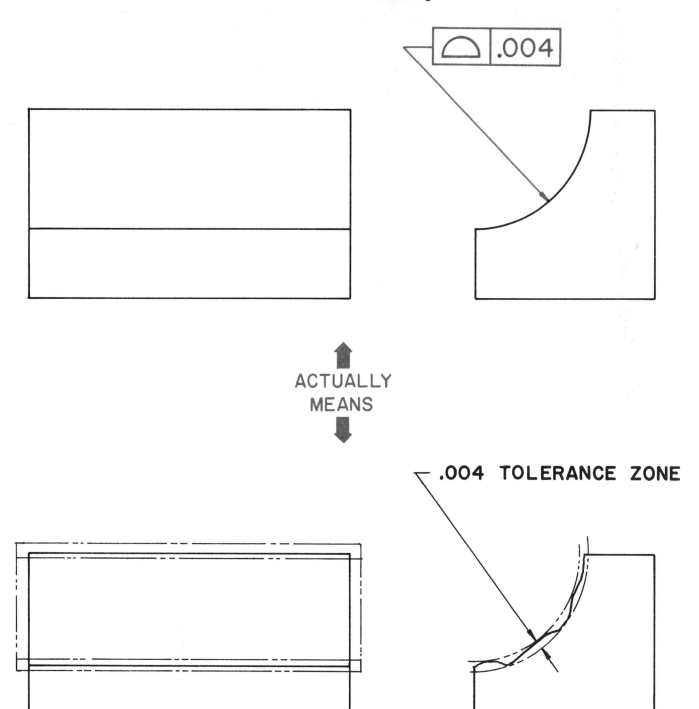

Fig. 12-24. Profile of a surface.

ORIENTATION TOLERANCES

Orientation tolerance symbols cover angularity, perpendicularity, and parallelism.

Angularity ⬚

ANGULARITY applies to a surface or axis which is at a specified angle (other than 90°) to a datum plane or axis. See Fig. 12-25.

Perpendicularity ⊥

PERPENDICULARITY requires a surface or axis to be at a right angle (90°) to a datum plane or axis, Fig. 12-26.

Parallelism ∥

PARALLELISM requires that a surface or axis remains the same distance at all points from a datum plane or axis, as shown in Fig. 12-27.

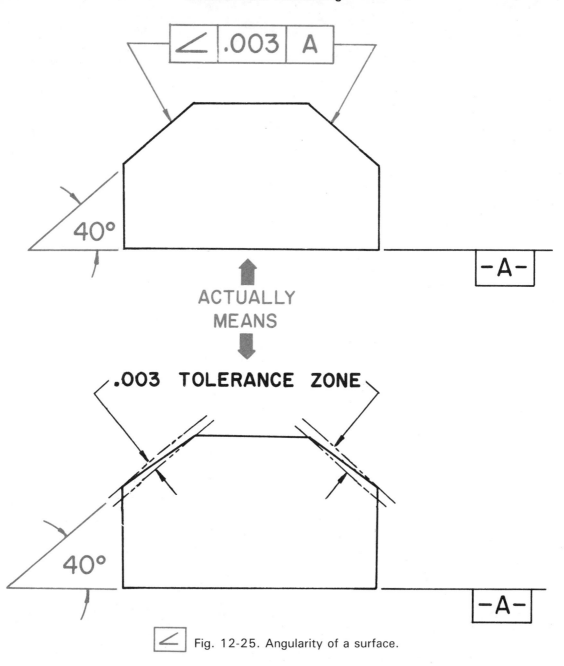

Fig. 12-25. Angularity of a surface.

LOCATIONAL TOLERANCES

Locational tolerance symbols include true position, concentricity, and symmetry.

True Position ⌖

TRUE POSITION defines a tolerance zone within which the axis or center plane of a feature is allowed to vary from the true (exact) position.

Coordinate tolerancing of a location provides a .020 inch square tolerance zone that results from coordinate plus and minus tolerancing. See views A and B in Fig. 12-28.

Locational tolerancing increases a tolerance zone by 57 percent more area by using a round tolerance zone. This occurs by placing the location dimension within rectangles, thus creating basic dimensions with exact values. The round tolerance zone, Fig. 12-29, is centered at the intersection of these basic dimensions.

Locational tolerancing is applied to a feature control, either on a MMC (maximum material condition), LMC (least material condition), or an RFS (regardless of feature size) basis.

MMC uses a modifier symbol Ⓜ. LMC uses a modifier symbol Ⓛ. RFS uses a modifier symbol Ⓢ. These modifiers should appear in the feature control symbol following the tolerance, as shown in Fig. 12-30.

MAXIMUM MATERIAL CONDITION (MMC) means that internal features such as holes and slots would be at their low limit (minimum) size, whereas

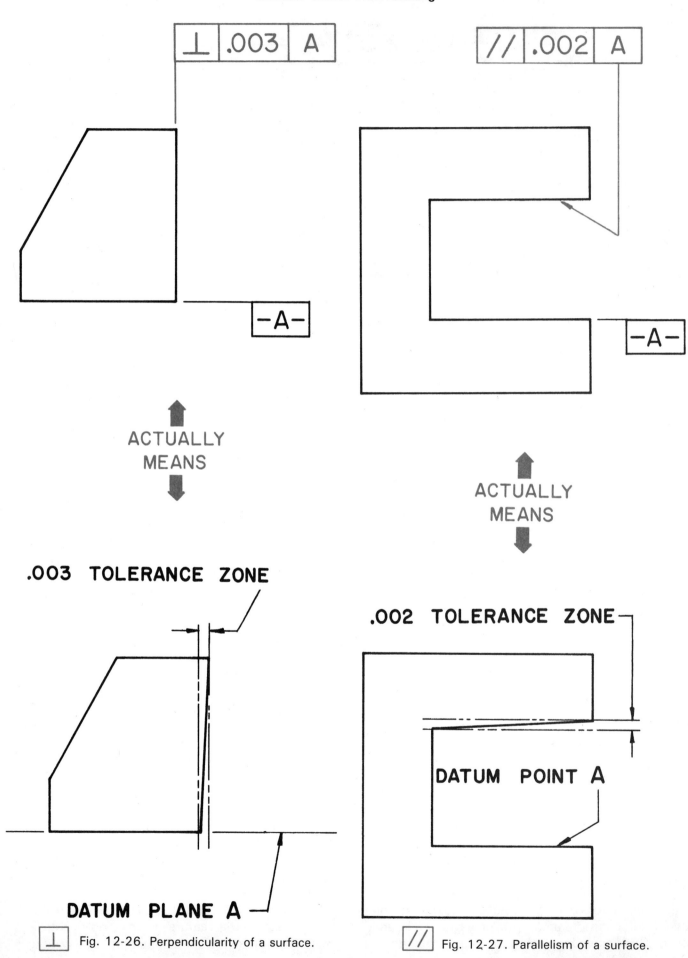

ACTUALLY MEANS

ACTUALLY MEANS

.003 TOLERANCE ZONE

.002 TOLERANCE ZONE

DATUM POINT A

DATUM PLANE A

⊥ Fig. 12-26. Perpendicularity of a surface.

// Fig. 12-27. Parallelism of a surface.

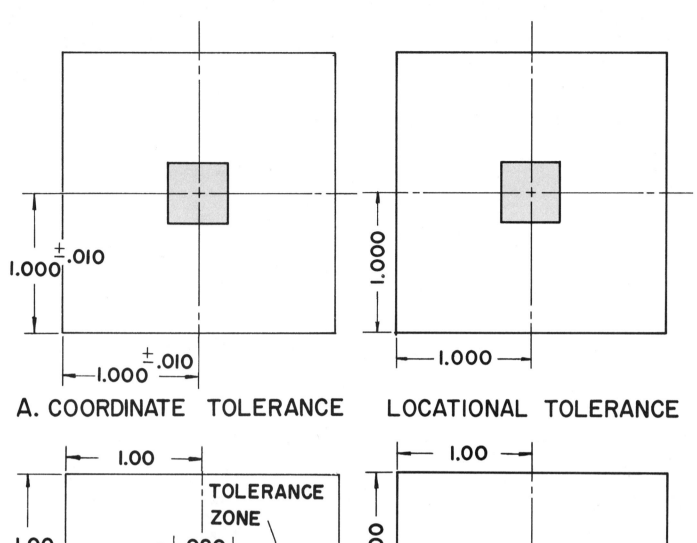

$1.000 \pm .010$

$1.000 \pm .010$

A. COORDINATE TOLERANCE

1.000

1.000

LOCATIONAL TOLERANCE

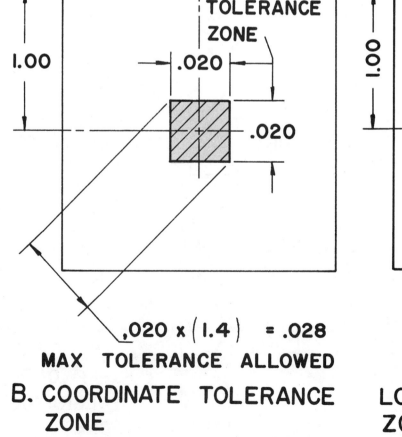

1.00

1.00

TOLERANCE ZONE

.020

.020

$.020 \times (1.4) = .028$

MAX TOLERANCE ALLOWED

B. COORDINATE TOLERANCE ZONE

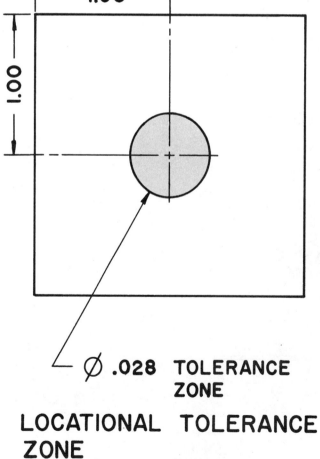

1.00

1.00

∅ .028 TOLERANCE ZONE

LOCATIONAL TOLERANCE ZONE

Fig. 12-28. Coordinate tolerance and tolerance zone.

Fig. 12-29. Locational tolerance and tolerance zone.

Fig. 12-30. Feature control symbol for locational tolerancing.

external features such as a shaft would be at their high limit (maximum) size.

EXAMPLE:
   MMC of a .750-.752 hole is .750 inch.
   MMC of a .750-.752 shaft is .752 inch.

Tolerance at MMC is interdependent on the feature size. If actual feature size has varied from MMC, the increase in the tolerance is allowed equal to that amount of variance.

Example:
   If a .010 inch tolerance zone is allowed for a .750-.752 inch hole at MMC, Fig. 12-31, the tolerance zone would increase to .011 inch for a .751 inch hole and .012 inch for a .752 inch hole. Tolerance increases are equal to amount of hole diameter change.

LEAST MATERIAL CONDITION (LMC) means that the internal features such as holes and slots would be at their high limit (maximum) size. Whereas, ex-

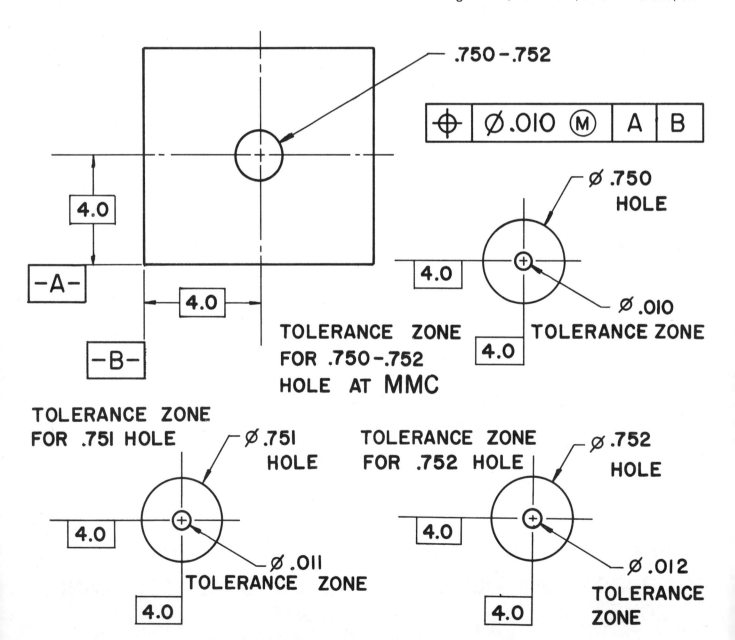

Fig. 12-31. Locational tolerance applied to a hole at MMC.

ternal features such as a shaft would be at their low limit (minimum) size.

EXAMPLE:
LMC of a .750-.752 hole is .752 inch.
LMC of a .750-.752 shaft is .750 inch.

Tolerance at LMC is also interdependent on the feature size. If actual feature size has varied from LMC, the increase in the tolerance is allowed equal to that amount of variance.

Example:
If a .010 inch tolerance zone is allowed for a .750-.752 hole at LMC, Fig. 12-32, the tolerance zone would increase to .011 inch for a .751 hole and .012 inch for a .750 hole. Tolerance change is equal to hole diameter change.

REGARDLESS OF FEATURE SIZE (RFS) means that the tolerance zone is limited to the specified value and does not change regardless of the actual size of the feature.

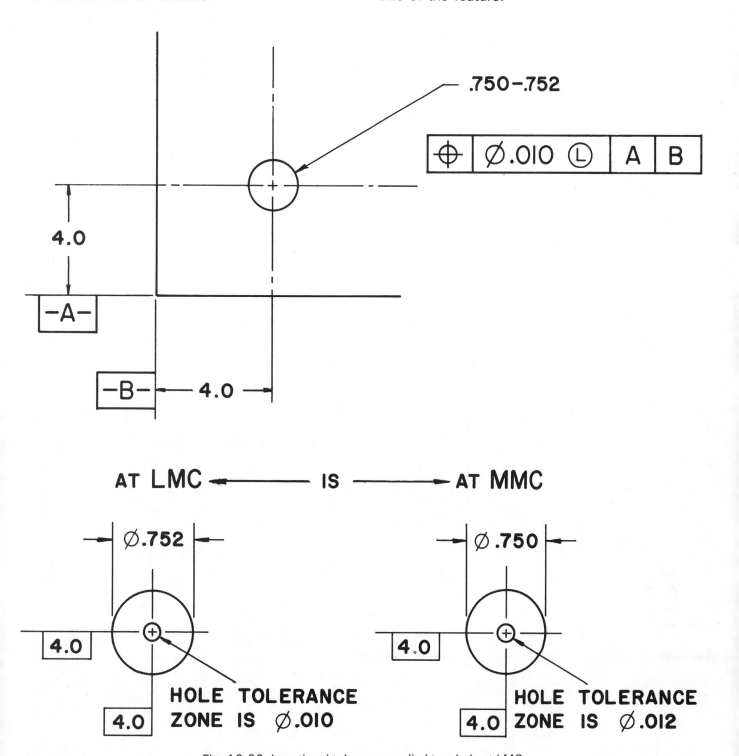

Fig. 12-32. Locational tolerance applied to a hole at LMC.

Example:
If a .010 inch tolerance is allowed for a .750-.752 hole the tolerance zone remains the same,

Fig. 12-33, regardless of whether the hole is .750, .751, or .752 inches in diameter. Tolerance at RFS is independent of feature size.

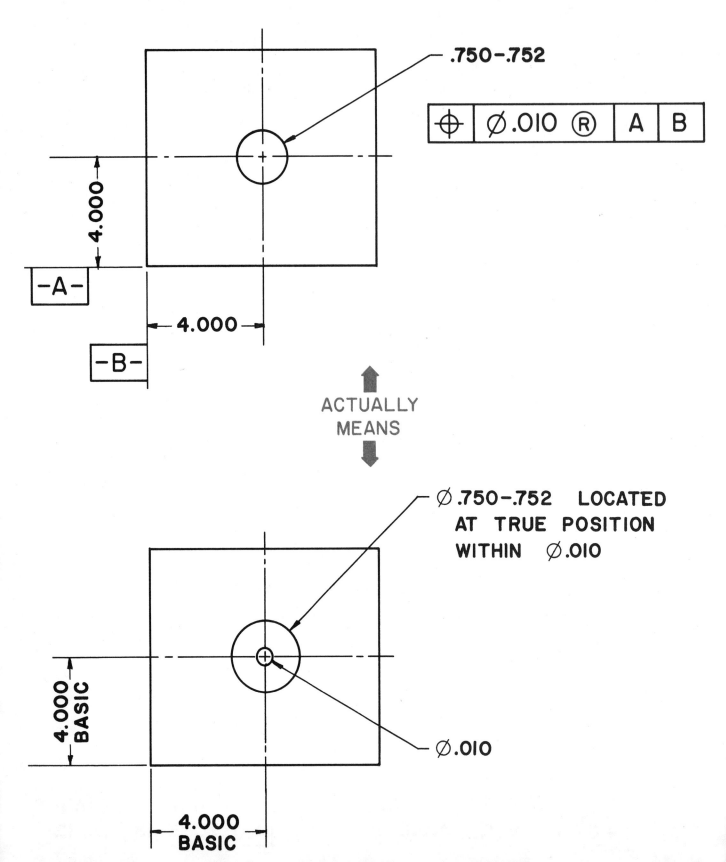

.750-.752

⊕ | ⌀.010 Ⓡ | A | B

-A-

-B-

4.000

4.000

ACTUALLY MEANS

⌀ .750-.752 LOCATED AT TRUE POSITION WITHIN ⌀.010

⌀.010

4.000 BASIC

4.000 BASIC

Fig. 12-33. Locational tolerance applied to a hole at RFS.

Concentricity

CONCENTRICITY exists when the axes of each part feature's surface of revolution are common to the datum axis. See Fig. 12-34.

Symmetry (use position symbol ⊕)

SYMMETRY exists when a part feature is symmetrically arranged about the center plane of a datum feature. See Fig. 12-35.

RUNOUT TOLERANCE

Runout tolerance symbols represent circular runout and total runout.

Circular Runout

CIRCULAR RUNOUT provides control of circular elements of a surface applied independently at any measuring location as the part is rotated 360 degrees. See Fig. 12-36.

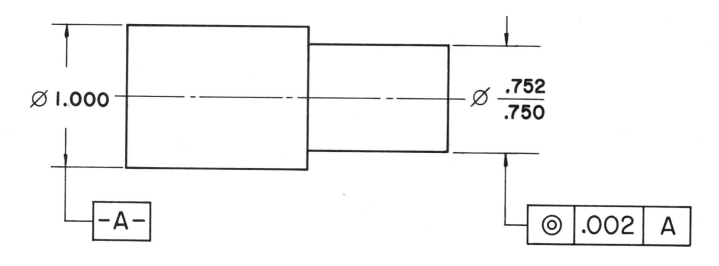

ACTUALLY
MEANS

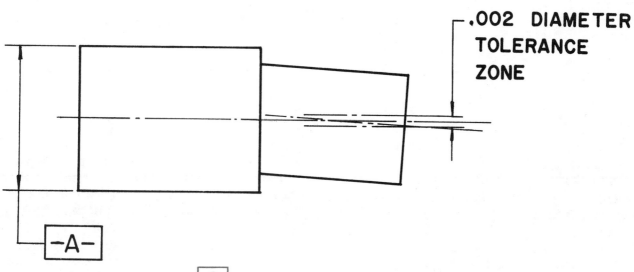

Fig. 12-34. Concentricity of a part.

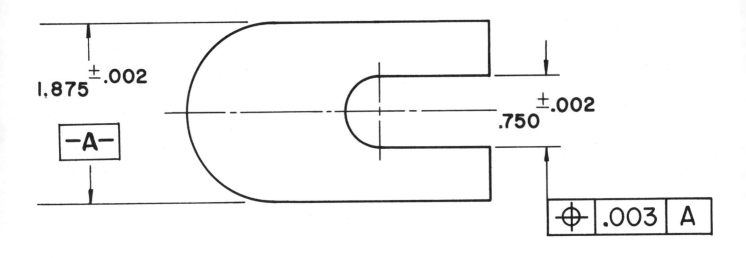

1.875 ±.002

-A-

.750 ±.002

⊕ | .003 | A

ACTUALLY
MEANS

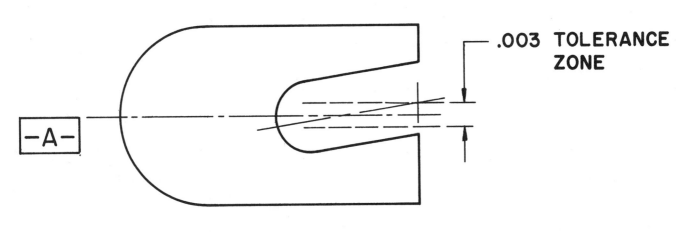

.003 TOLERANCE
ZONE

-A-

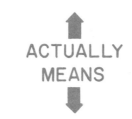 Fig. 12-35. Symmetry of a slotted part.

Total Runout ↗↙

TOTAL RUNOUT provides control of all circular elements of a surface applied simultaneously to all measuring locations as the part is rotated 360 degrees. See Fig. 12-37.

Runout is applied to surfaces at right angles to a datum axis and to surfaces existing around a datum axis.

Circular runout applied to surfaces existing around a datum axis controls cumulative form and locational characteristics (roundness, concentricity).

Circular runout applied to a surface at right angles to a datum axis controls wobble.

Total runout applied to surfaces existing around a datum axis controls cumulative form and locational characteristics such as roundness, concen-

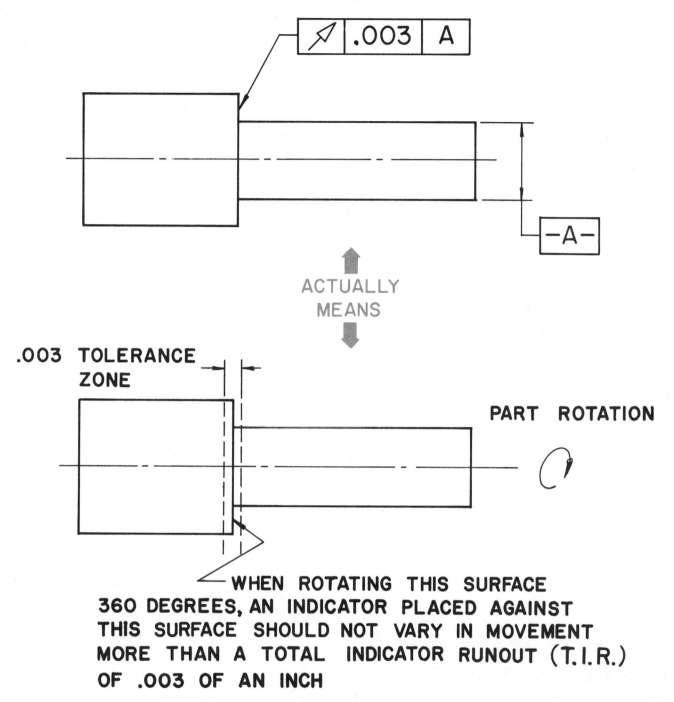

Fig. 12-36. Circular runout (wobble) on a shouldered surface.

tricity, angularity, straightness, taper, and profile of a surface.

Total runout applied to a surface at right angles to a datum axis controls wobble and flatness.

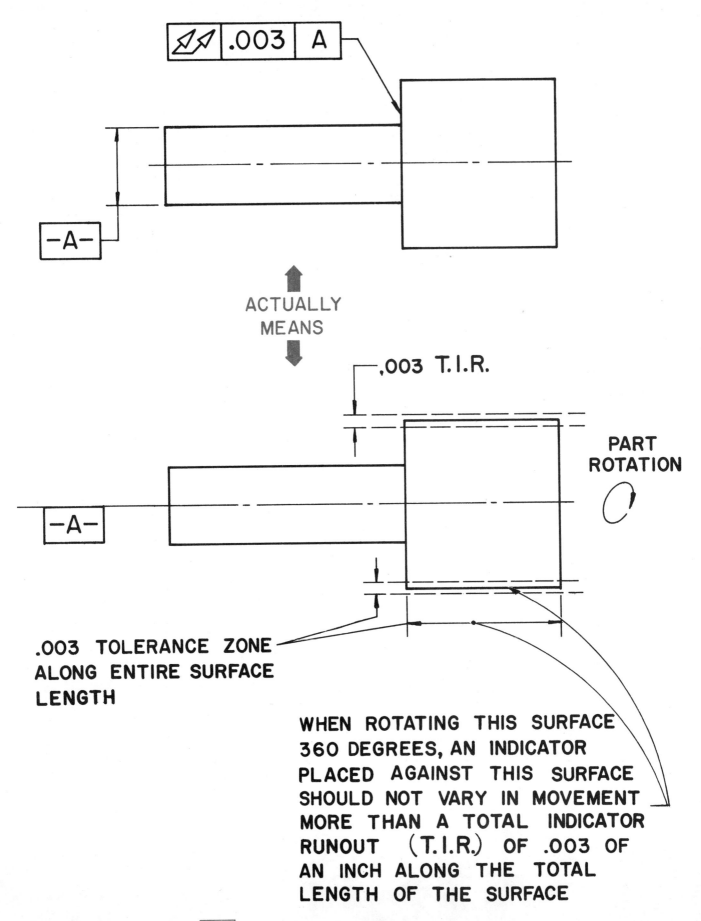

ACTUALLY
MEANS

.003 T.I.R.

PART
ROTATION

-A-

.003 TOLERANCE ZONE
ALONG ENTIRE SURFACE
LENGTH

WHEN ROTATING THIS SURFACE
360 DEGREES, AN INDICATOR
PLACED AGAINST THIS SURFACE
SHOULD NOT VARY IN MOVEMENT
MORE THAN A TOTAL INDICATOR
RUNOUT (T.I.R.) OF .003 OF
AN INCH ALONG THE TOTAL
LENGTH OF THE SURFACE

Fig. 12-37. Total runout on a shaft diameter.

DIRECTIONS—GEOMETRIC TOLERANCING QUIZ

Match each term or description with the correct symbol by placing the appropriate letter in the blank provided.

1. Datum identification symbol

2. Geometric symbol that specifies a tolerance zone between two concentric circles.

3. RFS

4. Geometric symbol denoting a tolerance zone between two concentric cylinders.

5. Symbol for counterbore

6. Geometric symbol used to identify a tolerance of location.

7. Datum target symbol

8. Depth

9. Geometric symbol creating a tolerance zone formed by two parallel planes.

10. Symbol requiring a surface or axis to be at right angle to a datum plane or axis.

11. Profile of a line

12. Symbol requiring axes of each part features' surface of revolution to be common to the datum axis.

13. Circular runout

14. Internal features (holes, slots) are at their low limit.

15. Parallelism

A	⊕ (datum target)
B	⌄
C	▱
D	-B-
E	⌯
F	⌰
G	Ⓛ
H	Ⓜ
I	◎
J	○
K	—
L	Ⓢ
M	⊕
N	⊥
O	⊔
P	∠
Q	//
R	⌒
S	⟋
T	⌓
U	⌀
V	□
W	⤓

1. _____

2. _____

3. _____

4. _____

5. _____

6. _____

7. _____

8. _____

9. _____

10. _____

11. _____

12. _____

13. _____

14. _____

15. _____

DIRECTIONS—POSITIONAL TOLERANCING QUIZ

Complete the chart for the part below by providing positional tolerance values for the hole shown as the machined hole size changes.

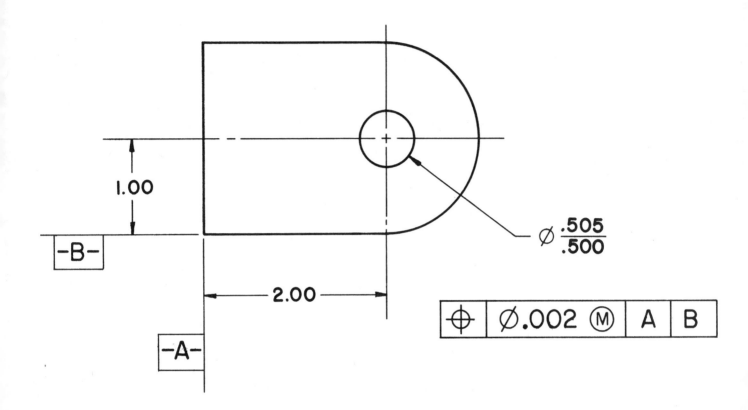

HOLE SIZE	POSITIONAL TOLERANCE
.505	
.504	
.503	
.502	
.501	
.500	

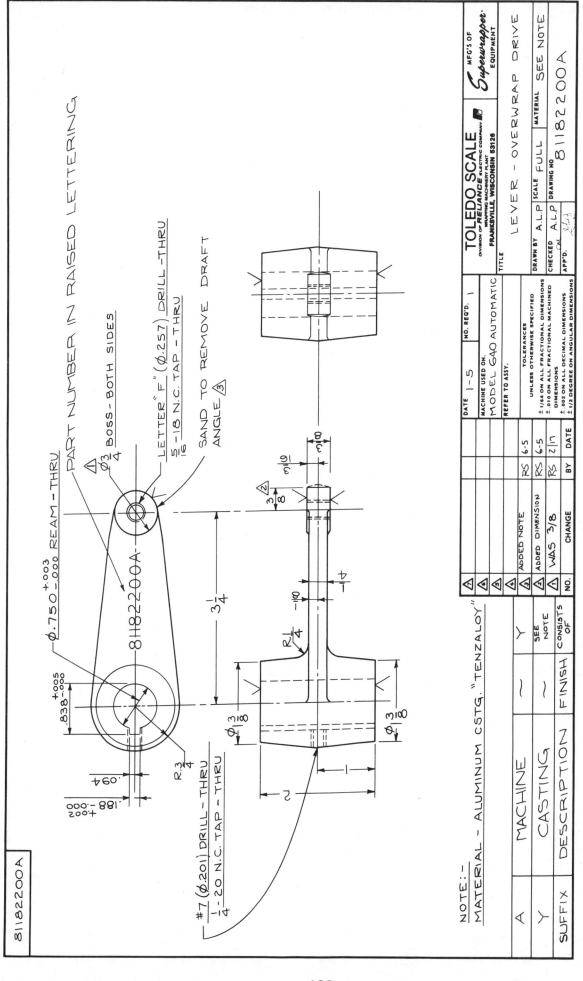

NOTE:-
MATERIAL - ALUMINUM CSTG. "TENZALOY"

SUFFIX	DESCRIPTION	FINISH	CONSISTS OF		
A	MACHINE	~	Y		
Y	CASTING	~	SEE NOTE		

NO.	CHANGE	BY	DATE		
△1	WAS 3/8	RS	7/11		
△2	ADDED DIMENSION	RS	6-5		
△3	ADDED NOTE	RS	6-5		

REFER TO ASSY.

MACHINE USED ON.
MODEL 640 AUTOMATIC

DATE 1-5   NO. REQ'D. 1

TOLERANCES
UNLESS OTHERWISE SPECIFIED
± 1/64 ON ALL FRACTIONAL DIMENSIONS
± .010 ON ALL FRACTIONAL MACHINED DIMENSIONS
± .002 ON ALL DECIMAL DIMENSIONS
± 1/2 DEGREE ON ANGULAR DIMENSIONS

MFG'S OF
Superwrapper
EQUIPMENT
TOLEDO SCALE
DIVISION OF RELIANCE ELECTRIC COMPANY
WRAPPING MACHINERY PLANT
FRANKSVILLE, WISCONSIN 53126

TITLE
LEVER - OVERWRAP DRIVE

DRAWN BY A.L.P	SCALE FULL	MATERIAL SEE NOTE
CHECKED A.L.P		DRAWING NO
APP'D.		81182200A

PART NUMBER IN RAISED LETTERING

△3 BOSS-BOTH SIDES

LETTER "F" (⌀.257) DRILL-THRU
5/16 - 18 N.C. TAP - THRU

SAND TO REMOVE DRAFT
ANGLE △3

⌀.750 +.003/-.000 REAM - THRU

81182200A

.838 +.005/-.000

.094

.188 +.002/-.000

R 3/4

#7 (⌀.201) DRILL - THRU
1/4 - 20 N.C. TAP - THRU

3 1/4

3/8

3/16

3/8

R 1/4

.100

1/4

⌀ 1 3/100

⌀ 3 1/100

2

1

Industry print of a multiview working drawing contains correct views, dimensions, part number, tolerances, and other specifications for manufacturing.

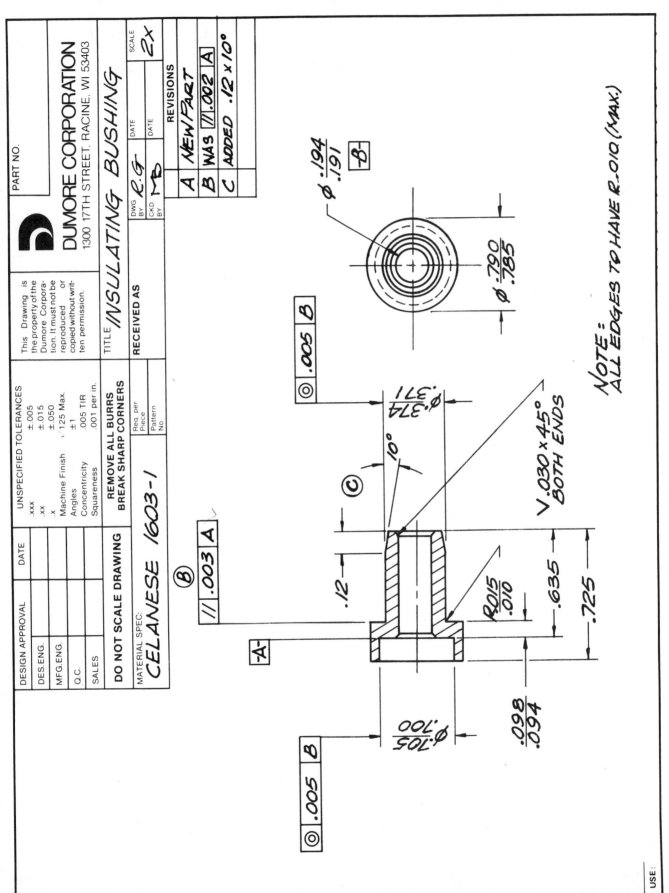

Fig. 12-38. Insulating Bushing.

DIRECTIONS—QUIZ QUESTIONS

1. The industrial prints in this section will test your print reading ability.
2. Study the views, dimensions, title block, and notes in Figs. 12-38 through 12-43.
3. Read the quiz questions, refer to the print, and write your answers in the blanks provided.

INSULATING BUSHING QUIZ

1. How deep is the counterbore?

1. _____

2. How long is the .790/.785 diameter?

2. _____

3. What is the high limit of the total length of the workpiece?

3. _____

4. Must the largest outside diameter be concentric to the counterbore diameter?

4. _____

5. State the size of the external chamfer.

5. _____

6. What feature is Datum B?

6. _____

7. What was revision C?

7. _____

8. What material is the bushing?

8. _____

9. What is the low limit of the large outside diameter?

9. _____

10. How long is the .374/.371 diameter?

10. _____

11. What is the approximate diameter of the second largest circle shown in the end view?

11. _____

12. What feature is Datum A?

12. _____

13. How far apart are the two surfaces that must be parallel?

13. _____

14. If the length of the .790/.785 diameter is at the high limit and the total length of the bushing is at its high limit, what would be the length of the .374/.371 diameter?

14. _____

15. What tolerance is allowed on concentricity?

15. _____

16. What finish is required on all surfaces?

16. _____

17. If the total length of the bushing is at low limit and the .635 length is at its high limit, what would be the counterbore depth?

17. _____

18. What tolerance is allowed on angles?

18. _____

19. What tolerance is allowed on two place decimals?

19. _____

20. What scale is the print?

20. _____

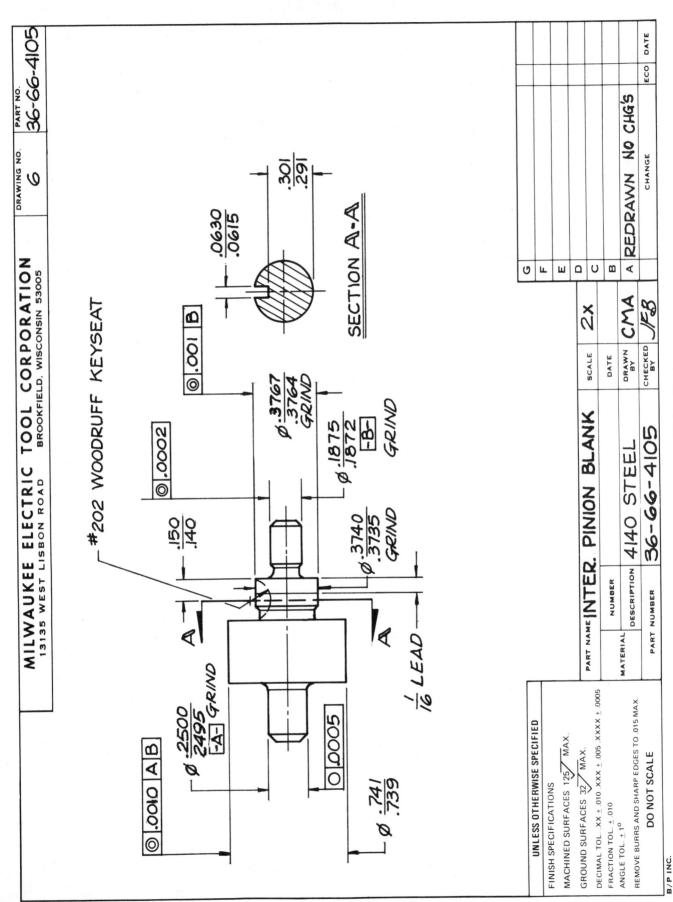

Fig. 12-39. Inter. Pinion Blank.

SECTION A-A

#202 WOODRUFF KEYSEAT

.0630 / .0615

.301 / .291

⌖ .001 B

⌖ .0002

.150 / .140

Ø .3767 / .3764 GRIND

Ø .1875 / .1872 GRIND -B-

Ø .3740 / .3735 GRIND

⌖ .0010 A B

Ø .2500 / .2495 GRIND -A-

⌖ .0005

Ø .741 / .739

A

A

1/16 LEAD

MILWAUKEE ELECTRIC TOOL CORPORATION
13135 WEST LISBON ROAD   BROOKFIELD, WISCONSIN 53005

DRAWING NO.	PART NO.
6	36-66-4105

G					
F					
E					
D					
C					
B					
A	REDRAWN NO CHG's				
	CHANGE		ECO	DATE

PART NAME **INTER. PINION BLANK**

SCALE	2x

	NUMBER	DATE	DRAWN BY	CMA
MATERIAL	DESCRIPTION **4140 STEEL**		CHECKED BY	JFB
	PART NUMBER **36-66-4105**			

UNLESS OTHERWISE SPECIFIED

FINISH SPECIFICATIONS
MACHINED SURFACES 125√ MAX.
GROUND SURFACES 32√ MAX.
DECIMAL TOL .XX ± .010 .XXX ± .005 .XXXX ± .0005
FRACTION TOL. ± .010
ANGLE TOL. ± 1°
REMOVE BURRS AND SHARP EDGES TO .015 MAX.

DO NOT SCALE

B/P INC.

INTER. PINION BLANK QUIZ

1. How wide is the keyway?

1. _____

2. List all the low limit dimensions shown for diameters.

2. _____

3. Why are length dimensions missing on this print?

3. _____

4. Why are datums not identified in the feature control frames for roundness?

4. _____

5. What dimension gives the location of the keyway?

5. _____

6. Which diameters are used as datums?

6. _____

7. What diameter woodruff key is used on this part?

7. _____

8. Do the views shown represent the actual size of the workpiece?

8. _____

9. What other dimensions besides lengths are missing on this print?

9. _____

10. What is the approximate depth of the keyseat?

10. _____

11. Which diameter requiring a roundness specification is more critical in tolerance allowed?

11. _____

12. What type of tolerance is concentricity?

12. _____

13. Which diameter must be concentric to the .3767/.3764 diameter?

13. _____

14. What tolerance is allowed on the smallest diameter shown on this print?

14. _____

15. Which diameter is shown in the sectional view?

15. _____

16. How many centerlines are shown on this print?

16. _____

17. What finish is required on ground surfaces?

17. _____

18. What is the size of the lead diameter?

18. _____

19. What type of material is used to make the part?

19. _____

20. What type of line is line A ?

20. _____

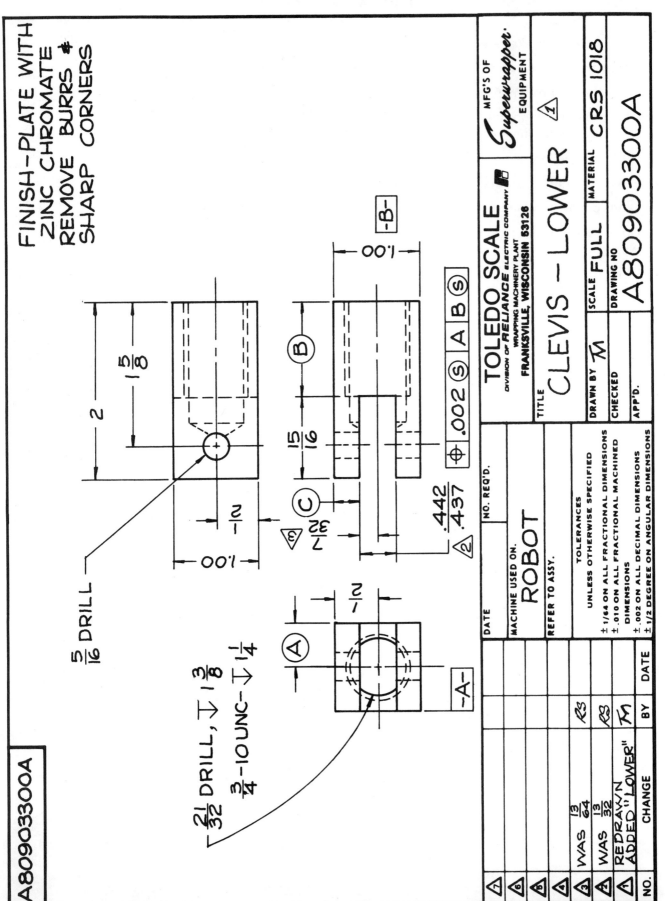

Fig. 12-40. Clevis—Lower.

CLEVIS-LOWER QUIZ

1. How far is the 5/16 drilled hole from the end of the workpiece?

2. What material is the workpiece?

3. What does (S) mean in the feature control frame?

4. How far can the centerline (plane) of the slot vary to the centerline of the part?

5. ⊕ is a locational tolerance symbol which refers to what part characteristic?

6. Datum feature B must have what relationship to datum feature A?

7. What is dimension (A) ?

8. How wide is the slot?

9. How deep is the 21/32 hole?

10. What feature is datum B?

11. What size tap drill is used on this print?

12. What was the original width of the slot?

13. How deep is the slot?

14. What is dimension (B) ?

15. What finish is required on this part?

16. What machine is used to make this part?

17. What is the high limit on the width of the part?

18. What is dimension (C) ?

19. Which dimension on the print was originally 13/64?

20. What thread series is the tapped hole?

1. _____

2. _____

3. _____

4. _____

5. _____

6. _____

7. _____

8. _____

9. _____

10. _____

11. _____

12. _____

13. _____

14. _____

15. _____

16. _____

17. _____

18. _____

19. _____

20. _____

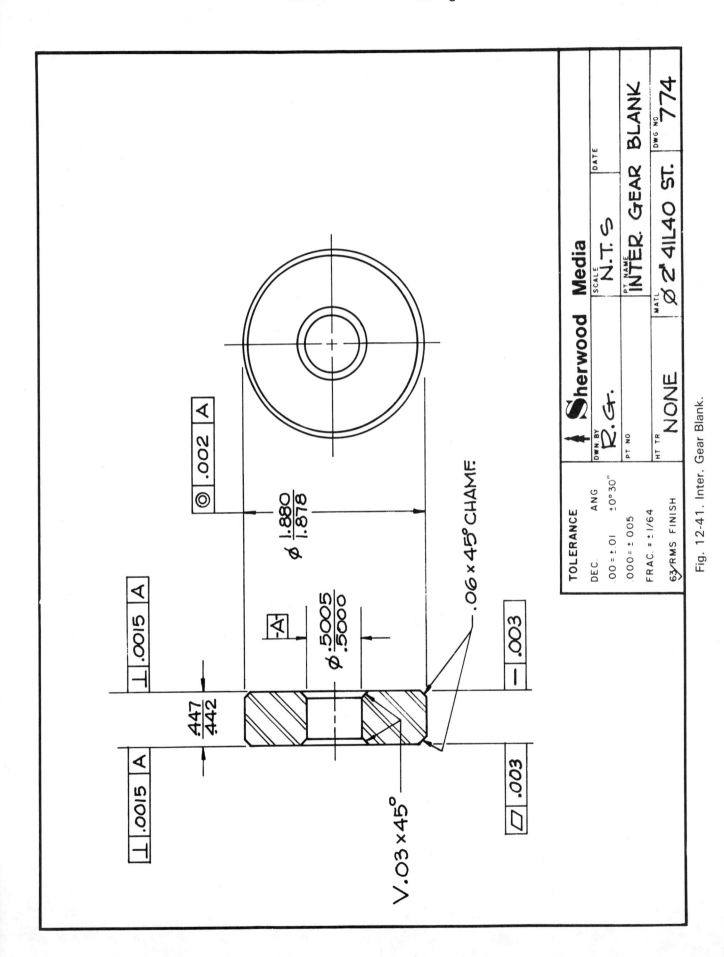

Fig. 12-41. Inter. Gear Blank.

INTER. GEAR BLANK QUIZ

1. What is the maximum thickness dimension allowed on the workpiece?

2. How many datums are given on the print?

3. What kind of material is used for the workpiece?

4. What finish is required on the ends of the gear blank?

5. What limits of flatness are given for the workpiece?

6. How concentric must the outside diameter be?

7. What feature is Datum A?

8. How many surfaces are given a tolerance for flatness?

9. What does the symbol before the value .03 × 45° represent?

10. What does the abbreviation N.T.S. mean?

11. Determine the tolerance allowed on the hole of the workpiece.

12. What does the symbol ⊥ mean?

13. Is this part heat treated?

14. What size chamfer is required for the workpiece?

15. Give an example of a feature control frame shown on this print.

1. _____

2. _____

3. _____

4. _____

5. _____

6. _____

7. _____

8. _____

9. _____

10. _____

11. _____

12. _____

13. _____

14. _____

15. _____

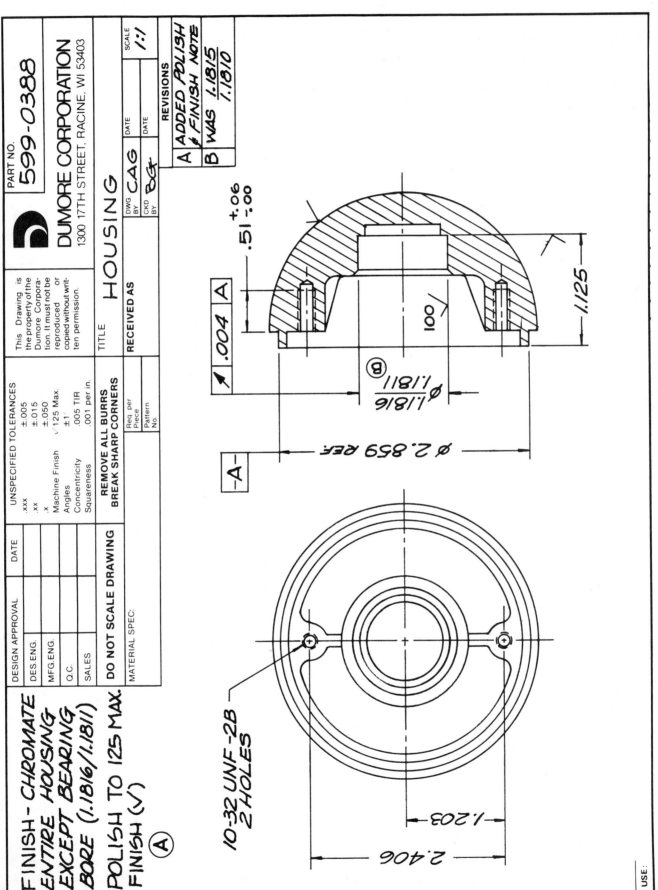

Fig. 12-42. Housing.

HOUSING QUIZ

1. How many tapped holes are in the part?

2. What dimensions are used in the location of the tapped holes?

3. What size is the bearing bore in the housing?

4. What type of view is the view to the right on the drawing?

5. What is the maximum depth allowed for the 10-32 UNF thread?

6. What is the minimum depth allowed for the 10-32 UNF thread?

7. How much tolerance is permitted on the bearing bore?

8. What kind of feature is Datum A?

9. What does the symbol ✦ represent?

10. Is the outside of the housing finished? If yes, to what finish?

11. What was revision B?

12. How much runout is permitted between the bore and the outside diameter?

13. How many ribs does this part possess?

14. How far apart are the tapped holes?

15. How is the housing finished?

1. _____

2. _____

3. _____

4. _____

5. _____

6. _____

7. _____

8. _____

9. _____

10. _____

11. _____

12. _____

13. _____

14. _____

15. _____

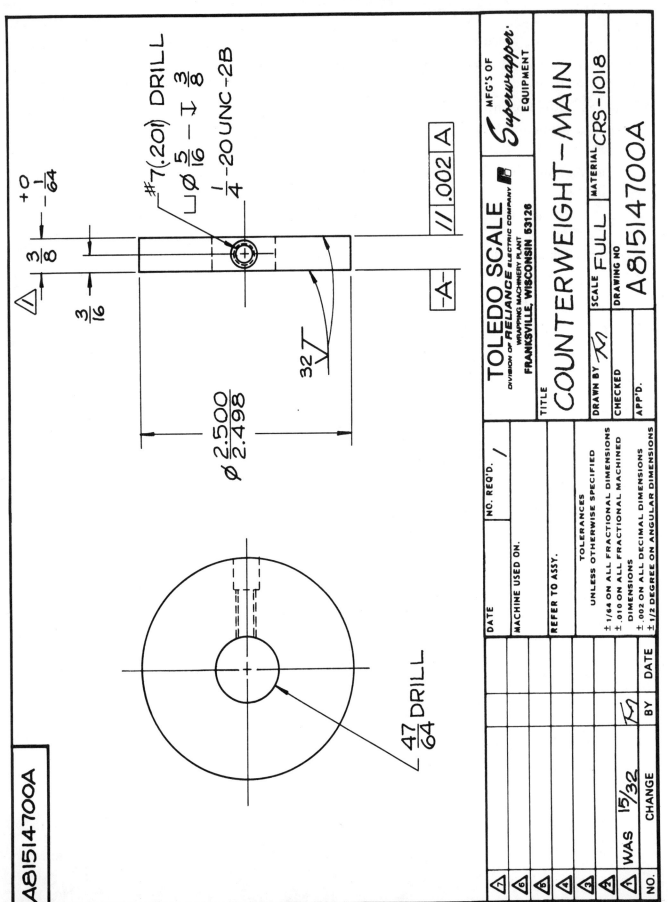

Fig. 12-43. Counterweight—Main.

COUNTERWEIGHT—MAIN QUIZ

1. What size is the counterbore diameter?

2. What is the minimum thickness allowed on the counterweight?

3. What tolerance is allowed on the outside diameter of the part?

4. How many datums are shown on the print?

5. What does the symbol ⊤ mean?

6. What does UNC mean?

7. How thick was the part originally?

8. What tolerance is allowed on the counterbore diameter?

9. How deep is the counterbore?

10. Approximately how long is the tapped hole?

11. What does the B mean in the thread designation?

12. How much wall thickness is there between the counterbore and the sides of the counterweight?

13. What type of fractional tolerance is shown on the print?

14. What type of material is 1018?

15. Is a #8 drill larger or smaller than a #7 drill?

1. _____

2. _____

3. _____

4. _____

5. _____

6. _____

7. _____

8. _____

9. _____

10. _____

11. _____

12. _____

13. _____

14. _____

15. _____

REFERENCE SECTION

DECIMAL EQUIVALENTS
Including Wire Gage, Letter, and Fractional Drill Sizes

Fractional Size Drills, Inches	Wire Gage Drills	Decimal Equivalent, Inches
	80	.0135
	79	.0145
1/64	..	.0156
	78	.0160
	77	.0180
	76	.0200
	75	.0210
	74	.0225
	73	.0240
	72	.0250
	71	.0260
	70	.0280
	69	.0292
	68	.0310
1/32	..	.0312
	67	.0320
	66	.0330
	65	.0350
	64	.0360
	63	.0370
	62	.0380
	61	.0390
	60	.0400
	59	.0410
	58	.0420
	57	.0430
	56	.0465
3/64	..	.0469
	55	.0520
	54	.0550
	53	.0595
1/16	..	.0625
	52	.0635
	51	.0670
	50	.0700
	49	.0730
	48	.0760
5/64	..	.0781
	47	.0785
	46	.0810
	45	.0820
	44	.0860
	43	.0890
	42	.0935
3/32	..	.0937
	41	.0960
	40	.0980
	39	.0995
	38	.1015
	37	.1040
	36	.1065
7/64	..	.1094
	35	.1100
	34	.1110
	33	.1130
	32	.1160
	31	.1200
1/8	..	.1250
	30	.1285
	29	.1360
	28	.1405

Fractional Size Drills, Inches	Wire Gage & Letter Drills	Decimal Equivalent Inches
9/64	..	.1406
	27	.1440
	26	.1470
	25	.1495
	24	.1520
	23	.1540
5/32	..	.1562
	22	.1570
	21	.1590
	20	.1610
	19	.1660
	18	.1695
11/64	..	.1719
	17	.1730
	16	.1770
	15	.1800
	14	.1820
	13	.1850
3/16	..	.1875
	12	.1890
	11	.1910
	10	.1935
	9	.1960
	8	.1990
	7	.2010
13/64	..	.2031
	6	.2040
	5	.2055
	4	.2090
	3	.2130
7/32	..	.2187
	2	.2210
	1	.2280
	A	.2340
15/64	..	.2344
	B	.2380
	C	.2420
	D	.2460
1/4	E	.2500
	F	.2570
	G	.2610
17/64	..	.2656
	H	.2660
	I	.2720
	J	.2770
	K	.2810
9/32	..	.2812
	L	.2900
	M	.2950
19/64	..	.2969
	N	.3020
5/16	..	.3125
	O	.3160
	P	.3230
21/64	..	.3281
	Q	.3320
	R	.3390
11/32	..	.3437
	S	.3480
	T	.3580

Fractional Size Drills, Inches	Letter Drills	Decimal Equivalent Inches
23/64	..	.3594
	U	.3680
3/8	..	.3750
	V	.3770
	W	.3860
25/64	..	.3906
	X	.3970
	Y	.4040
13/32	..	.4062
	Z	.4130
27/64	..	.4219
7/16	..	.4375
29/64	..	.4531
15/32	..	.4687
31/64	..	.4844
1/2	..	.5000
33/64	..	.5156
17/32	..	.5312
35/64	..	.5469
9/16	..	.5625
37/64	..	.5781
19/32	..	.5937
39/64	..	.6094
5/8	..	.6250
41/64	..	.6406
21/32	..	.6562
43/64	..	.6719
11/16	..	.6875
45/64	..	.7031
23/32	..	.7187
47/64	..	.7344
3/4	..	.7500
49/64	..	.7656
25/32	..	.7812
51/64	..	.7969
13/16	..	.8125
53/64	..	.8281
27/32	..	.8437
55/64	..	.8594
7/8	..	.8750
57/64	..	.8906
29/32	..	.9062
59/64	..	.9219
15/16	..	.9375
61/64	..	.9531
31/32	..	.9687
63/64	..	.9844
1	..	1.0000

TABLE OF CUTTING SPEEDS

Feet per Min.	30	40	50	60	70	80	90	100	110	120	130	140	150
Diameter Inches	REVOLUTIONS PER MINUTE												
1/16	1833	2445	3056	3667	4278	4889	5500	6111	6722	7334	7945	8556	9167
1/8	917	1222	1528	1833	2139	2445	2750	3056	3361	3667	3973	4278	4584
3/16	611	815	1019	1222	1426	1630	1833	2037	2241	2445	2648	2852	3056
1/4	458	611	764	917	1070	1222	1375	1528	1681	1833	1986	2139	2292
5/16	367	489	611	733	856	978	1100	1222	1345	1467	1589	1711	1833
3/8	306	407	509	611	713	815	917	1019	1120	1222	1324	1426	1528
7/16	262	349	437	524	611	698	786	873	960	1048	1135	1222	1310
1/2	229	306	382	458	535	611	688	764	840	917	993	1070	1146
5/8	183	244	306	367	428	489	550	611	672	733	794	856	917
3/4	153	203	255	306	357	407	458	509	560	611	662	713	764
7/8	131	175	218	262	306	349	393	436	480	524	568	611	655
1	115	153	191	229	267	306	344	382	420	458	497	535	573
1-1/8	102	136	170	204	238	272	306	340	373	407	441	475	509
1-1/4	92	122	153	183	214	244	275	306	336	367	397	428	458
1-3/8	83	111	139	167	194	222	250	278	306	333	361	389	417
1-1/2	76	102	127	153	178	204	229	255	280	306	331	357	382
1-5/8	70	94	117	141	165	188	212	235	259	282	306	329	353
1-3/4	65	87	109	131	153	175	196	218	240	262	284	306	327
1-7/8	61	81	102	122	143	163	183	204	224	244	265	306	327
2	57	76	95	115	134	153	172	191	210	229	248	267	287
2-1/4	51	68	85	102	119	136	153	170	187	204	221	238	255
2-1/2	46	61	76	92	107	122	137	153	168	183	199	214	229
2-3/4	42	56	69	83	97	111	125	139	153	167	181	194	208
3	38	51	64	76	89	102	115	127	140	153	166	178	191

STANDARD ABBREVIATIONS
USED IN DRAWINGS

A

ADD / Addendum
ADJ / Adjust
ALIGN / Alignment
ALLOW / Allowance
ALT / Alteration
ALUM / Aluminum
ALY / Alloy
ANL / Anneal
ANOD / Anodize
APPD / Approved
APPROX / Approximate
ASSY / Assembly
AUTO / Automatic
AUX / Auxiliary
AWG/ /American Wire Gage

B

BC / Bolt Circle

B/M / Bill of Material
BEV / Bevel
BHN / Brinnel Hardness Number
BNH / Burnish
BRG / Bearing
BRKT / Bracket
BRS / Brass
BRZ / Bronze
BRZG / Brazing
BUSH / Bushing

C

C to C / Center to Center
C'BORE / Counterbore
C'SINK / Countersink
CARB / Carburize
CDS / Cold-Drawn Steel
CH / Case Harden
CHAM / Chamfer

CI / Cast Iron
CIR / Circular
CIRC / Circumference
CL / Clearance
CONC / Concentric
COND / Condition
CONT / Control
COP / Copper
CPLG / Coupling
CR VAN / Chrome Vanadium
CRS / Cold-Rolled Steel
CSTG / Casting
CTD / Coated
CTR / Center
CTR / Contour
CYL / Cylinder

D

DAT / Datum

DCN / Drawing Change Notice
DF / Drop Forge
DIA / Diameter
DIAG / Diagonal
DIM / Dimension
DR / Drill
DWG / Drawing
DWL / Dowel

E

EA / Each
ECC / Eccentric
ECO / Engineering Change Order
EQ / Equal
EQUIV / Equivalent
EST / Estimate

F

FAB / Fabricate
FAO / Finish All Over
FIL / Fillet
FIM / Full Indicator Movement
FIN / Finish
FLG / Flange
FORG / Forging
FST / Forged Steel
FTG / Fitting
FURN / Furnish

G

GA / Gage
GALV / Galvanized
GRD / Grind
GSKT / Gasket

H

HCS / High Carbon Steel
HDN / Harden
HEX / Hexagon
HOR / Horizontal
HS / High Speed
HSG / Housing
HT TR / Heat Treat

I

ID / Inside Diameter
INSTL / Installation

K

KWY / Keyway

L

LAM / Laminate
LC / Low Carbon
LG / Length
LH / Left Hand

M

MACH / Machine
MAG / Magnesium
MATL / Material
MAX / Maximum
MECH / Mechanical
MI / Malleable Iron
MIL / Military
MIN / Minimum
MISC / Miscellaneous
MOD / Modification
MTG / Mounting

N

NO. / Number
NOM / Nominal
NORM / Normalize
NS / Nickel Steel
NTS / Not to Scale

O

OBS / Obsolete
OD / Outside Diameter

P

P / Pitch
PC / Piece
PROC / Process

Q

QTY / Quantity
QUAL / Quality

R

R / Radius
RD / Round
REF / Reference

REQD / Required
REV / Revision
RH / Right Hand
RH / Rockwell Hardness
RIV / Rivet

S

SCH / Schedule
SCR / Screw
SECT / Section
SEQ / Sequence
SERR / Serrate
SF / Spotface
SH / Sheet
SPEC / Specification
SPL / Special
SQ / Square
SST / Stainless Steel
STD / Standard
STK / Stock
STL / Steel
SYM / Symmetrical

T

TAP / Tapping
TEM / Temper
THD / Thread
THK / Thick
TIF / True Involute Form
TIR / Total Indicator Reading
TOL / Tolerance
TS / Tensile Strength
TS / Tool Steel
TYP / Typical

U

UNC / Unified Screw Thread
        Coarse
UHF / Unified Screw Thread
        Fine

V

VAR / Variable
VERT / Vertical

W

W / Width
WI / Wrought Iron
WT / Weight

INDEX

7388